PHILOSOPHIES OF QUALITATIVE RESEARCH

SERIES IN UNDERSTANDING STATISTICS
S. NATASHA BERETVAS   Series Editor

SERIES IN UNDERSTANDING MEASUREMENT
S. NATASHA BERETVAS   Series Editor

SERIES IN UNDERSTANDING QUALITATIVE RESEARCH
PATRICIA LEAVY   Series Editor

Understanding Statistics

Exploratory Factor Analysis
Leandre R. Fabrigar and
Duane T. Wegener

Validity and Validation
Catherine S. Taylor

Understanding Measurement

Item Response Theory
Christine DeMars

Reliability
Patrick Meyer

Understanding Qualitative Research

Autoethnography
Tony E. Adams, Stacy Holman Jones,
and Carolyn Ellis

Philosophies of Qualitative Research
Svend Brinkmann

Qualitative Interviewing
Svend Brinkmann

Video as Method
Anne M. Harris

Focus Group Discussions
Monique M. Hennink

The Internet
Christine Hine

Oral History
Patricia Leavy

*Using Think-Aloud Interviews and
Cognitive Labs in Educational Research*
Jacqueline P. Leighton

Qualitative Disaster Research
Brenda D. Phillips

Fundamentals of Qualitative Research
Johnny Saldaña

Duoethnography
Richard D. Sawyer and Joe Norris

*Analysis of the Cognitive Interview in
Questionnaire Design*
Gordon B. Willis

SVEND BRINKMANN

PHILOSOPHIES OF QUALITATIVE RESEARCH

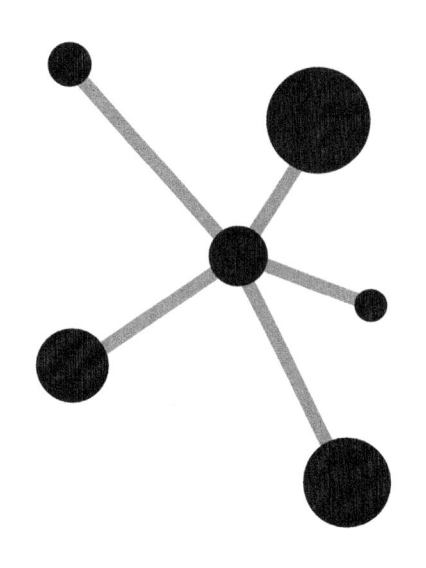

OXFORD
UNIVERSITY PRESS

OXFORD
UNIVERSITY PRESS

Oxford University Press is a department of the University of Oxford. It furthers the University's objective of excellence in research, scholarship, and education by publishing worldwide. Oxford is a registered trade mark of Oxford University Press in the UK and certain other countries.

Published in the United States of America by Oxford University Press
198 Madison Avenue, New York, NY 10016, United States of America.

© Oxford University Press 2018

All rights reserved. No part of this publication may be reproduced, stored in a retrieval system, or transmitted, in any form or by any means, without the prior permission in writing of Oxford University Press, or as expressly permitted by law, by license, or under terms agreed with the appropriate reproduction rights organization. Inquiries concerning reproduction outside the scope of the above should be sent to the Rights Department, Oxford University Press, at the address above.

You must not circulate this work in any other form and you must impose this same condition on any acquirer.

Library of Congress Cataloging-in-Publication Data
Names: Brinkmann, Svend, author.
Title: Philosophies of qualitative research / by Svend Brinkmann, Co-director, Center for Qualitative Studies, Professor of Psychology, Department of Communication and Psychology, University of Aalborg, Denmark.
Description: New York : Oxford University Press, [2018] |
Series: Series for understanding qualitative research |
Includes bibliographical references and index.
Identifiers: LCCN 2017012762 | ISBN 9780190247249 (alk. paper)
Subjects: LCSH: Qualitative research. | Social sciences—Philosophy.
Classification: LCC H62 .B6625 2017 | DDC 001.4/201—dc23
LC record available at https://lccn.loc.gov/2017012762

9 8 7 6 5 4 3 2 1
Printed by WebCom, Inc., Canada

CONTENTS

Preface vii
Acknowledgments xi

CHAPTER 1 Introduction: Philosophy and Qualitative Research 1

CHAPTER 2 The Historical Background: Philosophy from the Greeks to the 20th Century . . 21

CHAPTER 3 British Philosophies of Qualitative Research: Positivism and Realism . . . 47

CHAPTER 4 German Philosophies of Qualitative Research: Phenomenology and Hermeneutics 65

CHAPTER 5 American Philosophies of Qualitative Research: The Pragmatisms 91

CHAPTER 6 French Philosophies of Qualitative Research: Structuralism and Post-Structuralism 115

CHAPTER 7 Global Influences on Qualitative Research: New Philosophies 139

CHAPTER 8 Discussion 155

References 167
Index 179

PREFACE

This book is about the different philosophical paradigms and ideas that influence qualitative research. Its aim is to discuss and evaluate the ways that philosophical positions inform qualitative research as currently practiced. Unlike other contributions to the field, this book takes a historical perspective and shows how the philosophical ideas have evolved and influenced qualitative research in previous times and today. Today, qualitative researchers often report on their philosophical commitments (if they do so at all) in a separate section of their papers, but this book is written from the perspective that philosophical ideas influence everything in the research process from the first formulation of a research theme to the final reporting of the results. Therefore, it is preferable to highlight how this happens. Philosophy should thus not be thought of as a purely abstract discipline, disconnected from the practicalities of research, but rather as a concrete and pervasive aspect of all qualitative research practices. This book does not provide in-depth treatments of qualitative methods and techniques such as interviewing, document analysis, or participant observation, but rather aims to introduce and discuss the philosophical issues that are relevant regardless of the specific methods employed by qualitative researchers.

Overview of the Book

Chapter 1 provides some definitions of philosophy and qualitative research and gives an introduction to a number of philosophical themes that are relevant for qualitative researchers, covering the basics of epistemology and ontology and the various arguments in favor of realism and anti-realism. After this introductory chapter follow the seven main chapters of the book. These chapters build on each other to some extent, but they can also be read individually, depending on the reader's specific interests. Chapter 2 presents a selected history of Western philosophy from the Greeks to modern times, arguing that the very idea of qualitative research is a child of modernity's split between the objective and the subjective, *quanta* and *qualia*. This split became significant with the birth of modern natural science (Galilei, Newton, and Descartes), giving rise to the question of how to study those aspects of the world that do not seem to fit the perspective of the physical sciences. This question was answered in different ways by the British empiricists from John Locke onward and by Immanuel Kant in Germany. Chapter 2 is somewhat longer than the other chapters, and if the reader is eager to move ahead to the next chapters on paradigms that are specific to qualitative research, it can be skipped.

Chapters 3–7 present a number of different philosophies of qualitative research that have been important historically and continue to be so today. The different philosophies are introduced in relation to their respective birthplaces—four different birthplaces in total: the British traditions of positivism and realism (Chapter 3), the German traditions of phenomenology and hermeneutics (Chapter 4), the American traditions of pragmatism (Chapter 5), and the French traditions of structuralism and post-structuralism (Chapter 6). In each chapter, I first discuss the basic philosophical ideas before giving examples of how they have been applied in qualitative research practices. Next, in Chapter 7, I add a global perspective to the European and American traditions described in the previous chapters and include feminist, indigenous, and "new philosophies" from throughout the world. These philosophies are born in a globalized world and yet often stress the significance of local knowledges, and they typically cut across the philosophical traditions that have been introduced previously in the book. The five main chapters aim to locate and discuss the historical roots of the contemporary paradigms of qualitative research, which it is hoped complements other existing texts on philosophical approaches to qualitative research—for example, the chapter by Spencer, Pryce, and Walsh (2014), which lists postpositivism, social constructionism, critical theory, feminism, and queer theory. Unlike such other texts, the current book takes a deeper historical perspective and presents qualitative research today as intimately connected to various philosophical predecessors.

Chapter 8 summarizes and compares the different philosophies, and it constructs a matrix that includes all of them. I provide a brief discussion of how to "choose" a philosophical position as a qualitative researcher and whether this is a matter of choice at all (or rather a matter of one's basic view of humanity and the knowledge produced by humans).

ACKNOWLEDGMENTS

A big thank you to Patricia Leavy, editor of the *Understanding Qualitative Research* book series, for allowing me to write another volume after *Qualitative Interviewing*—and for patiently waiting for me to get this one done. Also thanks to students who throughout the years have listened patiently to my discussions about the role of philosophy in qualitative inquiry specifically and psychology and social science more broadly. I dedicate this book to you.

PHILOSOPHIES OF QUALITATIVE RESEARCH

1

INTRODUCTION

PHILOSOPHY AND QUALITATIVE RESEARCH

THE TERM PHILOSOPHY has become extremely broad, and there can be a philosophy of almost anything. Today, it is not unusual for the word philosophy to be used in connection with everything from large corporations ("Our philosophy is to...") to personal statements ("My personal philosophy is based on..."). The word can easily be misused, but it is commonly used to signal a serious reflection on fundamental principles. It literally means love (*philo*) of knowledge (*sophia*). When I was a philosophy student, my classmates and I used to joke that whenever someone is pressed to answer two or three "why" questions, that person will unavoidably begin philosophizing:

Let's go see a movie tonight!
Why?
Well, because we have worked hard and deserve do to something we simply feel like doing!
Why?
Well, I don't know.... Shouldn't we be allowed to do something we feel like doing from time to time? Uhm, perhaps it depends on what we feel like doing? How to distinguish worthy desires from unworthy ones?
OK, let's just go!

So, we see that after a few questions that press someone to go beyond the immediate—for example, an immediate desire, belief, or idea—we arrive at philosophical reflections. Anyone who is lovingly (*philo*) interested in the process of developing fundamental knowledge (*sophia*) can be considered a philosopher.

What Is Philosophy?

Today, philosophy is taught in university departments just like any other discipline such as medicine or geography. This means that philosophy has come to be thought of as a body of knowledge and as a set of analytic skills, which (many philosophers hope) are useful for other disciplines and sciences and perhaps even in the world of business and politics. This contemporary conception of philosophy as just another scholarly discipline, however, is closely and specifically connected to our modern world and its obsession with utility and instrumentalism. Philosophy must somehow be useful today if it is to be considered as a legitimate discipline. However, as the famous historian of philosophy Pierre Hadot argued in his classic book, *Philosophy as a Way of Life* (Hadot, 1995), this was very different in earlier ages, when philosophy was not such an abstract discipline but, instead, a form of life. According to Hadot, philosophy as originally conceived was not primarily characterized by intellectual abstractions but, rather, by its intentions to form people and transform their souls. This involved a kind of self-formation, or *paideia* in Greek, which was best cultivated through dialogue, as illustrated most famously in Plato's dialogues with Socrates as the key protagonist. The major schools of ancient philosophy such as Platonism, Aristotelianism, Cynicism, Epicureanism, and Stoicism all offered ways of becoming human and attaining a worthy life form. For most of the schools, the goal was to realize a common human nature, although there was no universal agreement concerning the exact character of this.

In the dialogue *Theaetetus*, Plato has Socrates state that "wonder is the only beginning of philosophy," which is probably the most famous quote on the nature of philosophy. Philosophy begins with wonder, Socrates says—that is, in a kind of puzzlement concerning existence—leading one to ask the fundamental questions about life and death, good and evil, knowledge and ignorance. For

Socrates—the philosopher who did not write anything but only asked questions—the practice of questioning was clearly more important than the answers being given. Often, in the Platonic dialogues, the conversationalists do not even arrive at an answer when discussing the perennial questions about justice, beauty, or the virtues but, rather, end in *aporia*—a state of confusion and intellectual impasse. But there is also in Plato a longing for eternally true and certain answers, especially concerning how to develop a good society, ultimately to be constructed on the basis of philosophers' universal knowledge of the true, the good, and the beautiful. Historians of philosophy still discuss the possible disagreements between Plato and Socrates on this point (Plato clearly being more in line with our contemporary theoretical understanding of philosophy, and Socrates representing the practice of questioning and philosophy as a process or an *elenchus*—a cooperative, dialogical method).

The contemporary philosopher Simon Critchley (2007) has given us an alternative conceptualization of the beginning of philosophy: "Philosophy begins in disappointment" (p. 1). On the one hand, as he sees it, philosophy stems from the disappointing realization that there is no justice in the world. We inhabit a violently unjust world, Critchley writes, where good people do not always win, and where tyrants may live happy lives. According to Critchley's analysis, the kind of disappointment that is associated with this realization has led to ethics and political philosophy and a quest to improve society. On the other hand, there is the disappointment that there is no God. Now, this is obviously a source of disagreement (both in the times when philosophy was born and also today), and most people in the world do probably believe in some kind of transcendent deity, but according to Critchley, *philosophy* has its origins in a reflection on what to make of the meaning of life in a universe without a God. In other words, how can we resist nihilism, the idea that everything is meaningless? Can there be meaning only if a metaphysical creature creates and supports it, or are we as finite, mortal creatures capable of constructing stable meaning in our lives? This age-old philosophical question is also a significant source of disagreement among qualitative researchers in a more tangible way: Is meaning something we *find* in our empirical materials or something we *make*? Are we primarily *explorers* of meaning or *creators* of meaning?

Whether it is Platonic wonder or Critchleyan disappointment that are singled out as existential origins of philosophy, it can be appreciated that the discipline was originally a practice concerned with human *life* and not just with abstract thought. Philosophy, in this original sense, is an existential affair, intent to help us live our lives, although this is too often forgotten in today's sterile and scientistic approaches to its teachings. The great American philosopher Stanley Cavell once defined philosophy as "education for grownups," which, in yet another way, points to the practical dimensions of the discipline. Humans have not invented philosophy simply in order to play intellectual games but, rather, because it is important for creatures like us to reflect on our world and find out how we ought to live. We shall see later in this book that the philosophical school known as pragmatism urges us to evaluate our thoughts and ideas relative to how well they enable us to live good lives.

A final aspect must be added to Critchley's point about disappointment: Fundamentally, philosophy is connected to the realization of human finitude. This was already evident in Socrates' words in the *Phaedo*, the dialogue that depicts the death of Socrates, when he explained that philosophizing is "training for death" (see also Hadot, 1995, p. 94). If we were immortal creatures, we would have no need to engage in philosophy because there would be no need to ask about what is truly significant (because nothing would then really matter, since everything could be done and redone endlessly). The idea of a connection between philosophy and mortality lived on through the teachings of such figures as Cicero in ancient Rome and Montaigne in French Renaissance humanism—philosophy being a sort of *Ars Moriendi*, a way of coming to terms with finitude in order to be able to live *and* die well, without fear or despair. With this, we are probably as far away as we can get from mainstream approaches to philosophy in current university departments. This book does not present philosophical exercises in any concrete way like the Stoics, for example, but it does share with Socrates, Cicero, Montaigne, and Critchley the view that philosophy is first and foremost a *practice*, a way of life that may inspire the qualitative researcher, just as it is also a systematic inquiry into the foundations of ethics, politics, aesthetics, and human knowledge. Addressing philosophies of qualitative research thus means addressing both the theoretical and the practical aspects of our inquiries and lives. I hope to be able to demonstrate that the philosophies of qualitative research are not just different sets of abstract principles but also different embodiments of modes of living and acting. Naturally, this leads to the next thorny question for a book on philosophies of qualitative research: What is qualitative research?

What Is Qualitative Research?

There are almost as many answers to this question as there are people who self-identify as qualitative researchers. One generic, but quite extensive, definition of qualitative research has been given by Norman Denzin and Yvonna Lincoln (2011a) in their authoritative handbook:

> *Qualitative research* is a situated activity that locates the observer in the world. Qualitative research consists of a set of interpretive, material practices that make the world visible. These practices transform the world. They turn the world into a series of representations, including fieldnotes, interviews, conversations, photographs, recordings, and memos to the self. At this level, qualitative research involves an interpretive, naturalistic approach to the world. This means that qualitative researchers study things in their natural settings, attempting to make sense of or interpret phenomena in terms of the meanings people bring to them. (p. 3)

I find this definition useful because it cuts across different paradigms in qualitative research (e.g., phenomenology, poststructuralism, and feminism), and it emphasizes the *practice* of research (rather than more abstract prescriptive principles), including how research should be seen as an activity in the world that changes it. However, like all definitions, it can be analyzed philosophically and be shown to embody various presuppositions that are not made explicit in the formulations as such. For example, the researcher is conceived as an *observer* who is located in the *world*. Many things can be implied by conceptualizing a researcher as an observer—does it mean, for example, that one is a passive spectator rather than an active participant in the activities of the world? And what is the world? This is definitely not an innocent word, but something laden with deep and potentially significant metaphysical assumptions. Is the world something that exists independently of the researcher/observer (that can be observed more or less veridically) or something that is created by the researcher in the act of observing? If there is one lesson to learn from the history of philosophy, it is that no concepts are innocent. Every term has a history and is saturated with philosophical presuppositions that may lead to various implications for research practice.

Another rich, yet potentially problematic term in the definition is "representation." What does it mean to say that researchers "turn the world into a series of representation"? What kind of connection can one imagine to exist between, for example, the sounds emanating from the mouth of an interviewee and the ink on paper representing these sounds—and their alleged meaning? And how, in turn, can an overarching analytic category created by the researcher when confronted with the many pages of transcripts be said to represent the meaningful content of a concrete utterance? Qualitative research involves long chains of transformation and translation from (staying with interviewing as an example) research questions to interview questions to interviewee answers to recordings to transcriptions to codings and categorizations, which ultimately end up becoming an analysis or interpretation that is published and (if one is lucky) read by people who might even use what they have read to change their opinions, actions, or policies (Brinkmann and Kvale (2015) take the reader through this whole process of interview research). But how to understand the relation between the signifiers (representations) and the signified (that which is represented)? This is an ancient philosophical question of representation that dates back to the Greeks and which has been given numerous different answers, most of which remain implicit when we talk about "representation" today. Finally, to refer to one last bit from Denzin and Lincoln's (2011a) definition, what can be meant by a "naturalistic approach," studying things in their "natural settings"? What would an "unnatural setting" be? The authors probably think of laboratories or other experimental settings, but why are these less natural than, for example, an interview encounter or even what goes on in a school or workplace when this is observed by a fieldworker? Is there an implicit philosophy of authenticity lurking in the background of the definition?

My goal here is not to tear the definition apart (as I stated previously, I find it useful and quite in line with how I would define qualitative research myself) but, rather, to argue that any definition imaginable (of qualitative research, but really of anything else too) implies a number of philosophical presuppositions about the researcher (as observer, participant, creator, interventionist, etc.), about the world (as dependent, independent, enacted, mute, etc.), and about the relation between them—and also about what can be known and about how it *ought* to be known. So we have all the major philosophical subdisciplines at work in an innocent-looking definition such as this one: ontology and metaphysics (What can be said to exist and how?), epistemology (What can we know and how?), ethics (What ought to be done and how?), and also—perhaps slightly more implicitly—political philosophy (What is a good and just society—and what role, if any, should research and academic analyses play in creating it?) and even aesthetics (What is beauty—and what is its relation to knowledge?). In the following section, I further discuss the various questions that philosophy deals with and why these are relevant in qualitative research practices. After that, I present an (impossibly quick and dirty) account of how these questions have been raised and answered from the Greeks until the emergence of modern philosophy in the early 17th century (one can only love the long historical perspective of a discipline that considers what happened in the 17th century "modern").

Basic Philosophical Questions

Philosophy deals with the fundamental questions for human beings—questions that are at once theoretical (e.g., What is the nature of reality?) and practical (e.g., How should we act?). It is commonplace to make a distinction between the theoretical subdisciplines of philosophy—to which ontology, metaphysics, and epistemology are normally said to belong—and the practical subdisciplines—with the main contenders being ethics and political philosophy. In addition, there are now a large number of specialized philosophical areas that are more difficult to place, such as the philosophy of mind, aesthetics, and the philosophy of law and jurisprudence.

Ontology and Metaphysics

Ontology and metaphysics are often mentioned together. The word "metaphysics" sounds like a New Age or spiritual term, but it simply means "after physics" in Greek, due to the simple fact that Aristotle's works on (what we now call) metaphysics were indexed after his work on physics. Physics is the science that deals with the physical world and how objects move through space and time, but metaphysics represents a deeper inquiry into what objects, space, and time ultimately are. Metaphysical questions thus concern the ultimate constituents of reality. Similarly, the term ontology derives from Greek and literally means the study (*logos*) of being (*ontos*). The battles in the history of philosophy over materialism (the world is ultimately made up of a physical substance, such as particles in movement) versus idealism (the world is ultimately constituted by ideas) seem to be perpetual, and in recent years they have been fought under the banner of the Science Wars, with materialists on the one side (defending the principles and methods of the natural sciences) and social constructionists on the other side (emphasizing the socially constructed nature of our ideas about the world).

More generally, this is also an ontological battle between realists and anti-realists. Realists believe that the world has certain properties independently of the human capacities of knowing them. There is a way the world is, which can be discovered and represented through the systematic use of scientific methods. Anti-realists deny this and argue that it is the human activity of knowing as such that creates what is known. This discussion can be staged in a quite general and overarching way, but typically it is framed in relation to specific regions of reality, so to speak. For example, most people are realists with respect to astronomical objects. We tend to think that the planet Neptune was there before it was discovered, and that it is not the investigative efforts of astronomists that have given this object its properties. These properties have been discovered rather than invented. However, when we speak of human life, matters are not that simple because the nature of human activity seems to be at least partly constituted by how we understand it and what we know about it. And what we know about it can seemingly be affected by the investigative efforts of human scientists. For example, it seems that a study of people's political attitudes already implies an interpretation of what is meant by "political" and "attitude," so these are not phenomena that exist independently and outside human interpretative practices. Furthermore, what we think of as "political attitudes" can sometimes be affected by the interpretations made by political scientists. This is the starting point for many interpretative researchers, who argue that the human and social sciences are trying to interpret a human reality that is already constituted by interpretations. Giddens (1976) has referred to this as a double hermeneutics—to indicate that social science consists of interpretations of interpretations—which may in turn become a triple hermeneutics, when researchers' interpretations feed back into the lives of those who are interpreted. The philosopher Ian Hacking (1995) has analyzed this process as a "looping effect," which can be said to be an interpretative dance between realism (there is *something* that is interpreted) and anti-realism (but this something is affected by our acts of interpretation). Hacking calls it "dynamic nominalism."

No qualitative research endeavor is without ontological commitments. Often, these are kept quite implicit and are perhaps even invisible to the researcher. But sometimes they are discussed explicitly. Many phenomenologists will say, for example, that their scientific goal is to discover the structure of human consciousness and being-in-the-world—to un-conceal it, to use an expression from Heidegger (1927/1962)—which reveals a commitment to a certain form of realism (which, of course, comes in many varieties). Many discourse analysts will say instead that they study how human talk itself brings the phenomena (which are talked about) into the world *in* the act of talking (Potter & Wetherell, 1987). Racism, for example, might from this perspective exist only as it is enacted in discursive practices, and researchers may themselves contribute to the development of or opposition to racism depending on how they address it. No talk is neutral, from the perspective of some forms of discourse analyses, because language is performative rather than representational—it creates what it talks about. This, obviously, is a form of ontological anti-realism. The discussion of realism and anti-realism dates back to the Greeks and was continued with particular force in the Middle Ages, and it is thus much older and broader than one could think.

Epistemology

Epistemology is the philosophical theory of knowledge. All qualitative research projects embody ideas about what knowledge is and how it can be obtained. Is knowledge something *found* or *seen* in the world—by mirroring reality as accurately as possible? This is the epistemological realist position. Or is knowledge *constructed* on the basis of different kinds of interest on behalf of the scientists? This is the anti-realist position, insisting that the mirrors created by scientists (their theories, concepts, and methodologies) are crucial for how something comes to be seen, according to the constructionist viewpoint. Or, to invoke a third perspective, is knowledge perhaps not at all about seeing? This could mean that neither the realist position that emphasizes the seen nor the anti-realist position that stresses the tools of seeing are helpful. Pragmatists, dating back to American philosophers such as Peirce, James, and Dewey in the late 19th century, argue that knowledge is about *doing* rather than *seeing*. Our knowledge is thus valid if it enables us to address and solve an important problem that arises in our individual or collective lives. Knowledge only exists in an act of knowing and not as a kind of object or representation of the world.

By stressing active doing rather than passive observation, the pragmatists deconstructed the modernist epistemological tradition from Descartes onward (discussed in more detail later). In contrast, epistemology was not a key philosophical subdiscipline for the ancient Greeks. The Greeks were more interested in the structure of reality than in how humans come to know this reality. But with Descartes in the early 17th century, philosophy was transformed, and epistemology became its prime discipline. As discussed later, this was prompted by the new natural sciences and their mechanical worldview, which created a split between the knowing subject and the objects of which the subject can know. And it was continued in the British empiricist tradition represented by Locke, Berkeley, and Hume in the 17th and 18th centuries.

In a famous essay, Taylor (1995) argued that we must now "overcome epistemology" because epistemology is so closely connected to this (false—in his view) notion of the isolated subject that knows the world only through its representations of the world. Taylor finds that we really cannot make use of this way of approaching humans and their knowledge. Not only pragmatists but also phenomenologists such as Merleau-Ponty (who stressed that we know the world through bodily engagement), Wittgenstein (who argued that the meaning of language arises from its practical use and not from representational properties), and other significant philosophers from the 20th century strove to develop other (non-epistemological) ways of understanding the human being and its relation to the world (Dreyfus & Taylor, 2015). These are presented in Chapter 2, but the key problem arising from the epistemological endeavor still persists in qualitative research: Is this kind of inquiry—based on descriptions and interpretations of people's sayings and doings—really about *representing* reality (as the epistemological tradition claims), or is it about something else for which we do not (yet) have a common term (different scholars use terms such as performing, enacting, or translating the world, depending on their philosophical leanings)?

Ethics

Moving on to practical philosophy and the question "How should I live?" we arrive at ethics or moral philosophy. "Ethics" derives from the Greek *ethos* (character) and "morality" from the Latin *mores* (which also means character, custom, or habit) (Annas, 2001). Although some philosophers stipulate a difference between the two terms, they can also be used interchangeably. Today, it is common to address ethics in relation to the social sciences generally, and qualitative inquiry specifically, simply as "research ethics." Ethics here becomes a set of principles that should guide responsible research practices but is then conceived as something external to the process of knowing itself. This is based on the idea that scientific knowledge is and ought to be value neutral. The legacy from positivism (discussed in Chapter 3) looms large here, and the idea is that science should be as objective, valid, and reliable as possible rather than being based on values. However, there are good reasons to believe that knowledge is always already laden with values. To begin by stating the obvious: The values held in high regard by positivists (as well as by many others), such as objectivity, validity, and reliability, are *also* values, which do not merely have an epistemic component (about finding truth) but also an ethical component (Brinkmann, 2005b). Being objective and reliable are indeed moral virtues that we value in people, whether they are engaged in research practices or their everyday lives.

Today, few scholars subscribe to the pure positivist view that science should be completely value neutral. If value neutrality is an illusion in any form of inquiry, the more interesting question becomes not *if* qualitative inquiry should build on values but, rather, *which* values researchers ought to subscribe to and possibly promote through their investigative endeavors. And here we find a rich variety of views and quite fruitful forms of disagreement. Some argue that qualitative research should both build upon and serve a democratic, communitarian ethic of care (Denzin & Lincoln, 2011a), whereas others argue that qualitative inquiry should build on the same kinds of value as other research practices (e.g., objectivity and validity) (Hammersley, 2008). There are today feminist, indigenous, social justice, queer, and critical race perspectives (and many others) that are each committed to laudable ethico-political values. The important point to understand in this discussion is that it seems impossible not to take sides, as Becker (1967) argued in his classic essay, "Whose Side Are We On?" That famous essay begins with the following sentence: "To have values or not to have values: The question is always with us" (p. 239). And, paradoxically, we must now say (against the positivists) that even the call to be "against values" in social science inquiry rests on certain values (for even value neutrality is a value), so we can never escape value-ladenness. Every fact loads a value, as Putnam (2002) has put it, and every value also loads a fact. But what *are* values? And how do we decide on which values we should be directed by in our qualitative research practices? These are key philosophical questions for qualitative researchers.

Political Philosophy

These questions also relate to the communities and societies within which qualitative research projects are conducted. This takes us to another subdiscipline of practical philosophy, namely political philosophy. The word "political" derives from the Greek term *polis*, which referred to the city state. Aristotle famously defined the human being as a *zoon politikon* (i.e., a political animal), which is a creature that can only realize its nature within the organized life of a community. The Greeks stated, *Polis andra didaske*, which means that the city is the teacher of man. This has been a foundational principle for the social sciences since Plato and Aristotle and continues today in all its varieties: We become what we are within organized social relationships.

Although I return to Aristotle in Chapter 2, it is relevant to briefly discuss in the present context how he conceived of (what we today call) the social sciences as a moral and political enterprise. It is impossible to overestimate the importance of Aristotle as a figure in the history of ideas. He is largely responsible for having divided up the sciences into their current forms, and his distinction between practical and theoretical sciences is fundamental. In Book II of his work on ethics, the *Nichomachean Ethics* (Aristotle, 1976), he distinguished practical inquiry from the theoretical:

> Since the branch of philosophy on which we are at present engaged is not, like the others, theoretical in its aim—because we are studying not to know what goodness is, but how to become good men, since otherwise it would be useless—we must apply our minds to the problem of how our actions should be performed. (p. 93)

We should be aware, Aristotle notes, "that any account of conduct must be stated in outline and not in precise detail [because] accounts are to be required only in such a form as befits its subject-matter" (p. 93). Aristotle saw it as a sign of an immature intellect to demand mathematical precision in practical matters, since the subject matter of mathematics can be grasped with theoretical precision, whereas this is hardly the case with respect to matters concerning human conduct. As he states later in his *Ethics* (p. 213), the political sciences—what we would call social sciences today—are species of what he called *phronesis* (practical wisdom) rather than *episteme* (theoretical knowledge of that which is of necessity) (see Table 1.1). The object of the practical sciences is human conduct, and this is a highly particularistic, versatile, and changing phenomenon. Practical knowledge about such things consequently "involves knowledge of particular facts, which become known from experience" (p. 215) rather than from theoretical deduction. In effect, here is an argument in favor of working with experiential, qualitative approaches in the study of human action in its concrete manifestations.

Table 1.1
Forms of Knowledge According to Aristotle

Form of knowledge and associated virtue	*Theoria—episteme*	*Poiesis—techne*	*Praxis—phronesis*
Paradigmatic example	Astronomy	Craftsmanship	Ethical conduct
Qualitative research orientation	Developing theoretically grounded knowledge (e.g., classical phenomenology)	Employing research techniques (e.g., interviewing as a craft)	Effectuating good actions (e.g., participatory action research)

Table 1.1 provides a simplified overview of Aristotle's concepts of basic kinds of knowing and the virtues humans must possess in order for them to know something well. First, we have *theoria* or theoretical knowledge, which concerns things that do not change in response to our activities of knowing (a paradigmatic example being astronomical knowledge). Aristotle uses the term *episteme* to refer to the virtue of thought needed to obtain theoretical knowledge. Next, Aristotle describes *poiesis* as a knowledgeable activity of making, which therefore concerns things that change in response to human activities. The paradigmatic example is the craftsman who possesses the adequate *techne* (a form of know-how) to build a house or make a pair of shoes. Finally, we have the activities of *praxis* and the virtue of *phronesis*, which resemble *poiesis/techne* in being about things that are affected by human action but differ because of the fact that activities of *praxis* are undertaken for their own sake and not for the sake of anything else. A craftsman builds something in order to make money or because it is useful to have a pair of shoes, but the *phronimos*—the person who is knowingly engaged in *praxis*—is doing something for its own sake. Ethical actions represent the paradigm of *phronesis* because good actions are conducted for their own sake and not

for the sake of any ulterior purpose (according to Aristotle, this also applies to theoretical or epistemic knowledge, which represents its own purpose). Aristotle also articulated other forms of knowledge, such as *sophia* (wisdom) and *nous* (the intellect), but the three forms laid out in Table 1.1 are the most significant ones. All three can be connected to qualitative research orientations—that is, to arguments about the basic form and function of qualitative inquiry. One can argue like classical phenomenologists, for example, that the basic task of qualitative research is to study the essential structures of consciousness and human experience. These can be understood theoretically through the use of phenomenology's methods, some scholars will argue. One can also argue, like Brinkmann and Kvale (2015), that the qualitative interview, to focus on one specific method, is a craft (a *techne*) that is ideally learned the way that (other) crafts are learned—that is, through apprenticeship and by observation, imitation, and practice. Finally, one can approach qualitative research from the point of view of *praxis* and *phronesis*. In recent years, there has been a resurgence of interest in such an Aristotelian view of the social sciences (as practical sciences). For example, in their sociological classic *Habits of the Heart*, Bellah, Madsen, Sullivan, Swidler, and Tipton (1985) developed what they called "social science as public philosophy." This perspective accentuated the fact that the researcher is within the society he or she is studying and also within one or more of its moral and political traditions (p. 303). Social science as public philosophy is political in the sense that it is part of the ongoing discussion of the meaning and value of our common life and in that it ideally engages the public. From an Aristotelian position, Flyvbjerg (2001) has argued that the social sciences are—or must become, if they want to matter—*phronetic*. Phronetic researchers place themselves within the political context being studied and focus on the values of the communities by asking three "value-rational" questions: Where are we going? Is this desirable? and What should be done? (p. 60). The raison d'être for the social sciences is, Flyvbjerg thinks, developing the value rationality of society—that is, enabling the public to reason better about values. This, I believe, is a very useful approach to the social sciences, but it leaves us again with the political philosophical question: What are values?

I have now introduced some themes and concepts from theoretical philosophy (ontology, metaphysics, and epistemology) and practical philosophy (ethics and political philosophy) and provided some examples of why and how these are very relevant for qualitative research practices. Some qualitative researchers place more emphasis on the former questions (being primarily interested in how their subject matter is constituted), whereas others are more interested in the latter normative questions about which values their research presupposes and implicates. Ideally, a philosophically well-grounded qualitative researcher is able to reflect on all these philosophical dimensions of his or her research practice, and how to do this in qualified ways is a key theme of this book.

Philosophically Informed Analytic Strategies in Qualitative Research

So far, in this chapter, I have provided various definitions of philosophy, and I have emphasized the rather more practical approaches to the discipline, as related to real, living human beings who are engaged in practices of knowing (which Hadot and others have called philosophy as a way of life). I have also introduced the basic subdisciplines of philosophy, and I discussed what qualitative research is. It is clear that no fixed definitions exist of either philosophy or qualitative research. The latter in particular can be approached from many different angles and involves a large number of different analytic strategies. As a way of organizing these different angles and strategies, it might be useful to introduce a distinction from Noblit and Hare's (1988) book on "meta-ethnography" (the art of synthesizing results from different qualitative studies) between different overarching analytic strategies in qualitative research (see Brinkmann, 2012). These strategies also represent significant approaches to qualitative research that are discussed in the following chapters, particular Chapters 3–6. They are the strategies of

- making the hidden obvious;
- making the obvious obvious; and
- making the obvious dubious.

In addition to these distinctions from Noblit and Hare, Bjerre (2015) has suggested a fourth analytic category of

- making the hidden dubious.

In order to provide some organization to the philosophies unfolded in the chapters of this book, I suggest that they—with a bit of goodwill—can be seen to incorporate these four general strategies or stances.

Making the hidden obvious is a critical stance that (in qualitative social science) grows out of a critique of positivism (see Chapter 3). It is articulated well within Marxism and critical realism. To criticize is here, for example, to uncover the hidden power structures that regulate human behaviors and influence the politics of human experience. Critical scholars talk about ideologies as systems of thought that work against people's interests, and the point of critical qualitative research is to see through the surface and demonstrate the ideological working mechanisms behind the manifest phenomena. Marx argued that all science would be superfluous if the outward appearance and the essence of things directly coincided (discussed in greater detail in Chapter 3). Science is here presented as having the goal of going beyond appearances to critically uncover a truer reality. Although Marx was a German, he worked in Britain for many years, and the whole contemporary movement of critical realism is primarily linked to the United Kingdom, which is why I refer to this as a British tradition in Chapter 3.

Making the obvious obvious is, in a very broad sense, a phenomenological stance that seeks to describe human life and experience (see Chapter 4). In colloquial terms, this stance has the goal of helping us see the trees as a forest and the forest as an array of trees. In qualitative research, this can be much more difficult than it seems, and I appreciate a quote by Pelias (2004) in this context, who argues that what I call a phenomenological stance is really the stance of poetry: "Science is the act of looking at a tree and seeing lumber. Poetry is the act of looking at a tree and seeing a tree" (p. 9). The poet, argues Pelias, is really the one who can see the world clearly as it is, which can be aided by the use of aesthetics (the term *aesthetic* originates from the Greek word for experience). The German traditions of phenomenology and hermeneutics placed emphasis on description and interpretation of human experience and are therefore, from a bird's-eye perspective, interested in making the obvious in our lives obvious for us, as will be shown in Chapter 4.

Making the obvious dubious is a more deconstructive stance that is introduced in Chapter 6 on the French traditions. It implies an attempt to question what we take for granted, not necessarily to uncover hidden mechanisms as in critical realist perspectives but, rather, to show that our meanings and understandings are unstable and endlessly ambiguous. This stance is found in French poststructuralist philosophies, which built upon but also criticized the idea of more stable structures of meaning found in structuralism. The concept of deconstruction was introduced by Derrida as a combination of "destruction" and "construction." It involves destructing one understanding of a text and opening for construction of other understandings (Norris, 1987). The focus is not on "what really happens" or "the real meaning of what was done" relative to some fixed structure but, rather, on presenting alternative versions of the real. It is affiliated with the "hermeneutics of suspicion" of the critical stance—that is, the idea that things may not be what they seem to be. But it does not search for any underlying genuine or stable meanings hidden beneath a text or an event. Meanings of words are conceived in relation to an infinite and ever-changing network of other words in a language.

Making the hidden dubious is a fourth stance introduced by Bjerre (2015). Unlike the previous stance that takes what we think is true and tries to show that it need not be, the fourth stance implies a critique of the very idea that there are hidden dimensions of meaning behind our common superficial experiences and practices in the first place. Bjerre mentions Nietzschean critiques of religion as examples of how the hidden has been made dubious. Interestingly, we can also mention positivism and its critique of metaphysics here: According to these perspectives, it simply does not make sense to talk about a transcendent realm of gods or ideas behind our lives as lived and experienced (see Chapter 3). Although it may seem a bit unfair (to some pragmatists), I believe it makes sense to say that pragmatism in general incarnates the stance of making the hidden dubious (see Chapter 5). For pragmatists, there are actions and consequences, and our scientific interpretations and explanations should be judged in light of the (beneficial) consequences they enable us to achieve—and not with reference to anything beyond or behind our practical lives. "Nothing is hidden" could be the slogan of this fourth stance, although it is in fact the title of a book about Wittgenstein's philosophy (Malcolm, 1988). But Wittgenstein and the pragmatists share the idea that language derives its meaning from the ways that concepts are used in practice—which is here for everyone to see in principle—and not from illusory hidden mechanisms or structures. This is addressed in Chapter 5 on American pragmatist traditions in qualitative research.

The critical, phenomenological, post-structural, and pragmatist stances are naturally not confined to their own birthplaces or traditions (I mean the traditions of critical theory and critical realism; phenomenology and hermeneutics; deconstruction; and pragmatism), although these schools of thought tend to cultivate such specific perspectives. Obviously, a phenomenologist can be critical and even deconstructive, and vice versa, so here I use the terms more generally to point to analytic strategies or stances that may be useful to bear in mind when one works as a philosophically informed qualitative researcher. The logic of one's study, and the way one presents and analyzes the material, may differ relative to one's interest in making the hidden obvious, the obvious obvious, the obvious dubious, or the hidden dubious. However, in Chapters 3–6, I deal with the four traditions in turn (each of which embodies several different traditions) relative to their geographical birthplaces: the British, the German, the French, and the American. But first, I present a brief overview of the history of philosophy in Chapter 2 to provide some historical context to the ensuing discussions.

2

THE HISTORICAL BACKGROUND

PHILOSOPHY FROM THE GREEKS TO THE 20TH CENTURY

AS I INDICATED in Chapter 1, the early history of philosophy is dominated by metaphysical questions, with epistemology gradually taking over from Descartes onwards in the 17th century. In the 20th century, some philosophers wanted the practical disciplines to gain primacy over both metaphysics and epistemology within philosophy. Pragmatists in the United States, for example, argued that all reasoning is basically practical reasoning about how to act (Garrison, 1999), and the moral phenomenologist Levinas (1969) argued that ethics should be thought of as "first philosophy"—but this is another story to be told later in the book. What matters in relation to the history of qualitative research is the fact that the idea of qualitative inquiry in many ways is a child of the split between the subjective and the objective that emerged with full force with the advent of the natural sciences and the corresponding philosophical reflections that were spawned by this. But let me rewind and begin with the earliest philosophers in the Western canon (and yes, this chapter limits itself to the philosophy of the imagined hemisphere we call the West)—in ancient Greece. In Chapter 7, I reflect critically on the possible ethnocentrism of philosophy.

Greek Philosophy

The first philosopher is conventionally said to be Thales. He claimed that "everything is water." Initially, this may strike one as a very odd thing to say, but perhaps it is not more mysterious than scientific statements such as "everything is atoms" or "all mental phenomena are brain processes." "Everything is water" can be considered a philosophical idea because it tries to say something about everything at the same time—outside the context of religion. The early philosophers in Greece, the pre-Socratics, were interested in the fundamental structure of the cosmos, but they disagreed concerning its basic principle: Thales claimed it was *water*, Anaximenes argued it was *air*, and Heraclitus rooted for *fire*. For the latter, the universe was in a state of flux and constant change, something which inspired philosophers such as Nietzsche and Deleuze millennia later. Democritus opted in his own way for *earth* as the defining principle of the world, giving rise to a theory of atoms that prefigured the theories that were to emerge in the much later era of modern science.

Regardless of the favored element—water, fire, air, or earth—a key discussion in early philosophy concerned permanence versus change. Heraclitus was the philosopher of change par excellence, arguing that being is an illusion as something permanent and static. For Heraclitus, becoming rather than being is the fundamental ontological principle. In opposition to this stood Parmenides, who believed that change is illusory because ultimately, he said, the world just *is* in a specific way. In the fragments that have remained from Parmenides, he posed the question, "How could what is perish? How could it have come to be? For if it came into being, it is not; nor is it if ever it is going to be." According to Parmenides, this meant that movement and change is an illusion. Only that which is, exists, and that which does not exist, is not. These seem to be self-evident truths that lead to the strange conclusion that nothing ever changes: For nothing can ever become what it is not, because that would mean that what is not in fact existed, which is faulty. Heraclitus and Parmenides stand opposed as two early giants of metaphysics with extreme viewpoints concerning permanence and change. Contemporary scholars still debate the metaphysical question whether there is anything permanent in the world (e.g., the laws of logic), or if everything is in flux, and also whether we can ever find permanent and universally valid knowledge about the world, or if we are left with ever-changing representations of a world in flux (this is a more epistemological question about our potentials for knowing).

Plato

It was Plato—arguably the most important philosopher in history, only rivaled by Aristotle and perhaps Immanuel Kant in the 18th century—who developed a synthesis of Parmenides' insistence on permanence and Heraclitus' emphasis on change. Plato developed a metaphysical theory according to which there is an eternal world of what he called ideas behind the ever-changing world of phenomena that we experience. So Plato did not deny that all our experiences involve change. Spring turns to summer, fall, and winter, and all living things grow before they turn old and ultimately perish. As illustrated in Plato's famous allegory of the cave, however, our perceptions of objects that move and change are merely akin to shadows on the wall of the cave, as copies of the true and unchangeable objects of the world. So, an individual horse is a changeable object, beginning its life as a foal before becoming a full-size horse whose body perhaps ends up as steaks on the table and its skin being used as leather to keep humans warm. Nothing about the animal as a tangible organism is permanent, but what makes the creature a horse, Plato argued, is not anything that can be observed by the senses in the world of change but, rather, is found in the way the horse takes part in the very *idea* of a horse. And this idea—like all ideas, including those of the good, the true, and the beautiful—is permanent and eternal. Otherwise, we would not know what we were talking about when talking about horses or anything else. Plato developed his arguments through his protagonist, Socrates, in the numerous deep and entertaining dialogues.

In the allegory of the cave, people who have hitherto only been exposed to the shadows on the wall can be brought to appreciate how these shadows are nothing but pale copies of a true reality by being freed from their chains and led outside the cave, ultimately to see how the sun is the source that lights up everything in our world. Plato compares this to how the idea of the Good (in his view) is fundamental among all ideas and is that which "lights up" the whole world. For Plato, as explained in the masterpiece

The Republic (Plato, 1987), only the philosopher—that is, the wise person who is in love with knowledge—can know about these matters, but the philosopher is easily blinded by having stared at the sun (true, eternal knowledge), which is why other people misunderstand him or her (well, him, given the patriarchy in Plato's times). Ideally, Plato argued, the philosopher, who has seen the true nature of reality, should be king in the human world and rule according to theoretical insights into what is true and good for human beings. Plato was no democrat, but wanted society to be based on true knowledge rather than contingent opinion. We here see a particular and quite influential view of the relationship between knowledge (e.g., as produced by qualitative researchers) and society: Scholars can discover the true and universal nature of the good society and the good life for people, which should then be implemented in practice. Today, this technocratic approach is widespread, with economists as perhaps the most visible scientists to inform policies, but also communities of positive psychology, for example, present their knowledge to governments about what a happy life is and how it can be brought about politically (for an analysis and critique, see Davies, 2015).

Plato's importance cannot be exaggerated. The philosopher Whitehead famously claimed in the 20th century that Western philosophy consists of footnotes to Plato. Especially when Plato's Greek ideas became part of Christian thought in the Middle Ages, a powerful synthesis of philosophy and religion emerged that was to dominate Western thought for centuries, in one way all the way up to Nietzsche. Like Christians, Plato seemingly believed in an immortal soul. The reason why we can obtain knowledge about the eternal ideas is because our souls come from the same eternal world and are only temporarily housed in the bodily shell. Thus, with Plato we have a dualist theory of body and soul (or mind) that later served as philosophical legitimation of much Christian thought.

Aristotle

Unlike Plato, his student Aristotle did not seem to believe in an immortal soul. For Aristotle, the soul is simply the form (in his technical language) of the matter of the living body. Thus, the soul is a name for the skills and disposition of a living creature, whether it is a plant with only a vegetative soul or a human being, which is alone in the known universe in having a soul with rational powers. There is a similar relationship between the body and the soul as there is between the eye and its ability to see: Seeing is the "soul" or the function of the eye. And just as the ability to see disappears when the eye perishes, so the soul disappears when the body dies. Also unlike Plato, Aristotle did not postulate eternal ideas (forms) behind the changing matter that we perceive. Form and matter always go together, and Aristotle can be called the first phenomenologist because he always began his inquiries with the phenomena at hand and not with anything behind or beyond what we experience (Nussbaum, 1986). The difference between Plato and Aristotle was captured in Rafael's famous fresco *The School of Athens* from 1509, on display in the Sistine Chapel, with the two philosophical giants occupying the center and with the other ancient philosophers around them: Plato pointing upward with one finger, signaling the transcendent nature of ultimate reality, and Aristotle extending his open hand (the palm facing down) toward the viewer and unto the world, symbolizing the earthly or immanent nature of reality. In many ways, Aristotle is much closer to contemporary qualitative researchers in the human and social sciences than Plato, which is why I shall go into some detail with his philosophy in what follows.

The *Nichomachean Ethics*, Aristotle's grand account of human beings and their social–ethical lives, begins by stating that by definition, all human activity aims at some good and cannot be understood in isolation from this good. Specific kinds of activity aim at specific kinds of end or good. The practices of medicine aim at health, military practices aim at victory, and economic practices aim at wealth (Aristotle, 1976, p. 63). But we do not, Aristotle observes, engage in all our activities in order to achieve something beyond or outside the activity itself (what follows builds on Brinkmann, 2011). We can say that I take piano lessons in order to learn the skill of playing the piano, but we normally cannot say that I play the piano in order to achieve something else. I simply play the piano because I like playing the piano. Those actions that are their own ends belong to *praxis* according to Aristotle, and the skill (or virtue) we need in order to perform a *praxis* well is called *phronesis* or practical wisdom, as shown in Table 1.1.

Actions considered as part of *praxis* are not instrumental or merely technical (*techne* in Greek), nor are they concerned with unchangeable objects like astronomy about which we may obtain true theoretical knowledge (*theoria*). Rather, they are conducted for their own sake and concern the world of changing matters. For something to count as a practice (*praxis*) for Aristotle, "it must be undertaken as a *constituent* means to *eudaimonia* (that is, the agent's reason must be expressible on these lines: 'Doing this is what, here and now, doing well is')" (McDowell, 1998, p. 7). *Eudaimonia* is a form of life (sometimes translated as "the flourishing life") that consists of actions and practices that we would choose for their own sake, and Aristotle defines the *eudaemon*—the person who lives well—as "one who is active in accordance with complete virtue, and who is adequately furnished with external goods, and that not for some unspecified period but throughout a complete life" (Aristotle, 1976, p. 84). By "virtue," Aristotle meant a disposition to act and feel in the right manner as demanded by the circumstances. Acting and feeling in the right manner has no exterior purpose but is its own end. Due to the kind of creature that humans are, we have a natural function, and the virtues are those capacities and dispositions that enable us to realize our function and live a form of life that is good and proper for humans—and thus be *eudaemon*. Studying this whole field of the good life (*eudaimonia*), human practices (*praxis*), and our embodied skills (*phronesis*) was Aristotle's idea of what we call the human and social sciences today, and indeed this was a qualitative rather than quantitative inquiry.

But Aristotle was much more objectivist than most of today's qualitative researchers. He believed that the human being was defined by its function—just as organs in the body are defined by their functions. So, if we succeed in discovering the function of, for example, the heart, then we know what a heart essentially and objectively is (viz. a blood pump, although this was discovered only long after Aristotle by William Harvey). Likewise, the *Nichomachean Ethics* is about the function of the human being, which Aristotle defines in two ways: Both ways, however, are related to our capacity to reason because reason is the distinct mark of the human, according to Aristotle. We are rational animals with the capacity for conceptual thought and thus with the capacity to respond to meanings and reasons. We are creatures that are able to give reasons for what we do and expect such reasons in return from others. So the human function is the exercise of our rational and discursive powers, which, of course, have to be cultivated in and by the communities of which we are a part. Aristotle here emphasizes two kinds of rationality, which is something that has confused many exegetes. On the one hand, his *Ethics* culminates in the final Book 10 with the assertion that *contemplative* activity is the highest form of activity, something we do for its own sake, and thus is constitutive of *eudaimonia*. So the perfected kind of life is one that involves the ability to theoretically contemplate our place in the cosmos, something that contributes to nothing else but the act of contemplation itself. In the preceding sections of the *Ethics*, however, much more emphasis is put on the practical and social aspects of life, on our capacity for practical rationality in a community of practical reasoners, and some have even wondered whether the final Book 10 could be an addition made by one of Aristotle's students. For in the other parts of the *Ethics*, "political" activities are at center stage. "Political" derives from the Greek word for city, *polis*, and Aristotle famously claimed that the human being is a political animal, a *zoon politikon*—an animal that can live a proper form of life only in the socially organized context of a community. Like some qualitative researchers today (e.g., Flyvbjerg, 2001), Aristotle would probably have said that the practice of human and social science is intimately connected to the political community of which it is a part.

If we are to evaluate Aristotle's importance for the kinds of questions that qualitative researchers ask, we can say that, on the one hand, he seems more relevant than ever (Harré (1997) has even argued that we should move forward to Aristotle!): Aristotle depicted the human being as a thoroughly social being—much in line with contemporary social constructionists—that can only live fully within the community of a *polis*. Aristotle also argued that in order to understand human action, we cannot simply seek the mechanical causes behind human behaviors. He made a distinction between four kinds of causes that are relevant when considering what happens in the world: the material cause (that which makes up something), the formal cause (the form of something), the efficient cause (that which effectuates a change), and the final cause (the end toward which something tends). In modern philosophy, after the breakthrough of the scientific worldview, this pluralism of causes came to be reduced to the efficient cause alone: The cause of what happens is a previous state of affairs and how this affects something as described by the natural laws. But, for Aristotle, this was just one causal aspect, and when acting human beings are the subject matter, it is the final cause—the reasons of our actions— that is of utmost importance if we are to identify the action. This would later be repeated by hermeneutical thinkers in the 19th century (e.g., Dilthey, 1894/1977) and social constructionists in the 20th century—that we need to *understand* the reasons and motives behind our actions and not just *explain* our behaviors with reference to mechanical causal laws.

On the other hand, there is something about Aristotle that is difficult to swallow for most human and social scientists today, at least for those who identify as qualitative researchers: He believed that there are objective facts about human nature and about what humans ought to do in their lives. Thus, for Aristotle, there is no problem with the otherwise infamous is/ought dichotomy, which later came to play such a huge role, especially with David Hume in the 18th century. The reason is that values (the oughts) are a certain kind of fact in the Aristotelian scheme, namely facts about the proper functions of things, including human beings. This means that it becomes possible to state objectively what a good human being is (just as it is possible—given the Aristotelian worldview— to state objectively what a good heart is, namely one that performs its function of blood-pumping well). Aristotle was a realist and a naturalist concerning values; he argued that values and other significant qualities are "out there" to be discovered by "qualitative researchers." But this sits very uneasily not only with the mechanical view of nature that arose after the Renaissance but also with the anti-essentialist philosophies that came to reign from the latter half of the 20th century onward. These are much more relativist and perhaps subjectivist than the old Greek philosopher.

Summarizing the Greek philosophers as a whole, we can say that Greek philosophy began with a metaphysical speculation about the ultimate constituents of the universe, leading to Plato's notion of two worlds—one of eternal ideas and another of experienced phenomena—and also to Aristotle's (in my view) more contemporary philosophy, which was a kind of phenomenology that aimed to understand the phenomena of human life. In Aristotle's teleological (*telos* means "purpose" in Greek) worldview, all things are defined by their purposes, and these purposes are to be found out there in the world. For Aristotle, meaning, value, and purpose are a part of the world, and they are not merely the subjective projections of human beings unto a value-neutral reality. In other words, the qualities studied by qualitative researchers (how humans throughout the world think, feel, and act) are just as objective as other properties studied by the sciences. However, the sort of teleological explanation that Aristotle deemed necessary in order to understand human beings—as creatures that can act for a reason—was exorcised from the sciences by Galileo and Newton, as discussed later, and this created some of the important background to current discussions in qualitative research.

Medieval Philosophy

Between the Greeks and the modern scientific breakthrough, however, is a long period of medieval philosophy, which represents much more than "dark ages" in human thought. The representation of the Middle Ages as particularly "dark" was in many ways a construction brought about by the later Enlightenment philosophers, who needed a contrast to their own modern emphases on science and progress. In reality, medieval philosophy was a flourishing and at times quite open and tolerant intellectual enterprise, institutionalized in monasteries in which scholars sought to unite the Greek ideas with Christian theology. I have already mentioned how this happened with Plato's philosophy (with St. Augustine being an important early Platonist, Christian philosopher, and church father), which suited the Christian doctrine of an immortal soul very well, but also Aristotle's philosophy (which came to Europe through Arab philosophers) was synthesized with Christianity by influential philosophers, the most famous of these being Thomas Aquinas, more or less the official philosopher of the Catholic Church.

This book is not a broad history of philosophy but, rather, is focused on the relevance of philosophical discussions for qualitative researchers today, and for that reason I must limit myself to treat just one significant discussion from medieval philosophy, which has come to be known as the medieval problem of universals. In a way, this discussion rehearsed a disagreement between Plato and Aristotle, but in a very sophisticated and poignant way. The problem concerns our concepts or categories: In what way do these refer or relate to objects in the world? Our concepts are universal ("horse," "depression," "social class," etc. refer to a large number of things), and yet we seem to only encounter particulars in our dealings with the world. Even the most perfect circle we can draw will never be perfectly circular—at least if we zoom in to the atomic level—so it seems that there is a gap between the particular and imperfect instances of things, on the one hand, and the universal concepts as such, on the other hand. So what are the universals? Do they even exist? And how should we conceive of the relation between the particular and the universal? These questions caused a heated debate for centuries in the Middle Ages, and it continues to preoccupy philosophers and researchers today, although few of these are aware that they more or less re-enact a discussion that is very old. Broadly speaking, the medieval philosophers gave three different answers.

The *Platonic* answer was defended by Anselm of Canterbury among others, and it can be called *idealism* because it argues that the universal ideas (concepts or forms) are primary and in a sense exist before the concrete particulars. According to this viewpoint, universals thus have a kind of independent existence, and although this is probably a rare theory in the human and social sciences today, it is often defended by mathematicians, for example, who argue that the objects they deal with in their inquiries are ideal and detached from particular things. For many mathematicians, numbers are real in the sense of belonging to a non-empirical, ideal realm.

Next is the *Aristotelian* theory, sometimes called realism, which claims that universals are real but always instantiated in concrete particulars. This was Thomas Aquinas' solution. Universals are neither "before" nor "after" the particulars but, rather, always "in" them. Whenever we find something that we can categorize (e.g., "horse" or "depression"), we understand the given something *as* the something it is only because it is imbued with a concept. And we understand the ideal only because it is instantiated in a particular object. The ideal is found *in* the real.

Finally, we have the *nominalist* theory, according to which universals are not actually real at all, at least not in a sense in which they are independent of human minds and can be discovered by our investigative efforts. Rather, humans *choose* to group things together, and it is the very act of choosing this which creates the seemingly universal categories and not anything preexisting in the things themselves. *Nomen* means "name" in Latin and indicates that it is the name that constructs the category rather than something in the world that forces us to use a specific category. The world is carved up into different classes of things by our very act of naming these classes. In medieval philosophy, it was especially William of Ockham who defended this theory. He is best known today for what we call "Ockham's razor," which is a principle that states that one should always choose the simplest hypothesis— that is, the one with fewest assumptions—whenever competing hypotheses about something exist. So, if there is no need to postulate universals existing "out there" (either ideally or in the things themselves), we should abstain from postulating them at all, if we wish to explain the human capacity for conceptual thought and the formulation of categories.

Modern readers, who have read philosophy of science or followed some of the contemporary debates between realists and constructionists, can easily recognize these medieval discussions in present-day guises. We still have many intellectual battles between universalists on the one side, arguing that (qualitative) researchers can and must (if they are to be called scientists) discover the essences of the mind and social life, and anti-realists or nominalists on the other side, arguing that general categories are constructed by the human mind rather than something found in the world. In between, perhaps, we have various schools that defend some kind of interaction between the knowers and the known—for example, dialectical materialists and perhaps critical realists, who can be said to represent an updated Aristotelian position. Thus, the discussion between Platonists, Aristotelians, and nominalists is still going on today, even if the current discussion partners may not always be aware of the ancient roots of their disagreements.

Renaissance Philosophy

It is not easy to say precisely when medieval philosophy was superseded by the Renaissance. But something happened from the 14th and 15th centuries onward. A whole cultural movement began, which we now know as the Renaissance, because it sought to rehabilitate (i.e., to give new birth to) Greek ideas. Observational science and naturalistic art began to flourish, particularly in Italy with famous painters such as Leonardo da Vinci and Michelangelo, writers such as Dante, and thinkers such as Pico della Mirandola, a neo-Platonist nobleman who published the *Oration on the Dignity of Man* in 1486, setting the stage for a new form of humanism. The human being was now put on center stage—and this was a self-confident, dignified, and perceptive human being, prepared for the individualism that was to affect Western thought and societies for centuries to come.

Leonardo is a wonderful example of "the Renaissance man," witnessed in his many-sided intellectual endeavors. Leonardo examined the world through his paintings (the *Mona Lisa* is arguably the most famous painting in the world) and detailed sketches (e.g., of the human body), and he invented machines and technologies that still amaze us today. Renaissance painters invented perspective in the modern sense, and this in itself is an indication of the shift from Greek and medieval cosmologies. In the earlier religious cultures, it was God who was conceived as the center of the universe, and this was reflected in the ancient works of art. Regardless of where the depicted objects are located, they are typically of the same size because God sees everything at the same time from "a view from nowhere" to use the phrase made famous by Thomas Nagel (1986). It is not yet the particular human viewpoint that is important. This changes, however, with the Renaissance, where the person viewing the painting becomes the center. Thus, objects become depicted relative to the observing individual and not in relation to an all-encompassing deity. This whole development can be viewed as a move away from the theocentric cosmology (with God as center) that was characteristic of metaphysically oriented philosophies and toward an anthropocentric cosmology (with man as center) that was to become significant, and which also indicates a shift to epistemology as the prime philosophical discipline, as I have argued. Beginning in the early 20th century, the anthropocentric cosmology is then again surpassed by a development of what some call a polycentric cosmology (Qvortrup, 2003) in which there is no one center but, rather, multiple centers or nodal points from which one may approach the world. This is also known as perspectivism and postmodernism, emphasizing "local knowledges" (in the plural) as discussed later in the book, and it was reflected in the art forms from the 20th century such as expressionism—for example, in the celebrated paintings by Picasso that depict a human being from several angles at the same time. There is no longer a privileged perspective on the world but, rather, multiple perspectives and thus many truths. But this takes me ahead of the story, so let us return to the Renaissance humanists and ask how they influenced what we now know as qualitative research.

Beginning in northern Italy in the 14th century, scholars began to study their contemporary society in comparison with the previous Roman empire, and this historical awareness initiated the discussion about "what in human affairs is owed to nature and what to custom" (Smith, 1997, p. 86). This was very important in legal scholarship, which "established patterns of thought that became the social sciences." (p. 86). Some Renaissance scholars, such as Hugo Grotius, followed in the footsteps of Thomas Aquinas and argued that there is a natural law, a pattern and order in the natural world, which ideally is reflected in the positive law that governs human action. But what we call social science now was originally closely connected to practical sciences such as law. This has also been discussed by the philosopher Stephen Toulmin (2001), who made a distinction between the concept of "rationality" (to designate later modernity's pure and abstract reason as seen in the sciences) and the concept of "reasonableness" (referring to our "impure," practical reason). Toulmin's project traces the philosophy of practical reason (already studied as *phronesis* by Aristotle) back to the humanists, and he argues that it is no less reasonable than the theoretical variant. Toulmin shows how practical reason in particular was an object of inquiry for the Renaissance humanists of the 16th century, who lived and wrote prior to the great modern breakthrough, which was heralded by the Galilean, Cartesian, and Newtonian elevation of general, eternal, and theoretical truths. In contrast, humanist writers such as Montaigne, Erasmus, and Shakespeare developed a philosophy of life that promoted concrete, temporal, and practical truths. For Toulmin, in a postmodern world today, we should try "to recapture the practical modesty of the humanists, which let them live free of anxiety, despite uncertainty, ambiguity, and pluralism" (Toulmin, 1990, p. 105), which is one indication of how the Renaissance humanists may prove to be relevant today. Toulmin wants us to consider the humanists' teachings as a way to live without angst in an uncertain, ambiguous, and pluralistic world. Renaissance humanism points the way to the concrete, temporal, and practical dimensions as an incubator of human certainty, knowledge, and ethics.

In the philosophy that developed in Europe from the Renaissance onward, humanism was represented by figures such as Erasmus of Rotterdam and Michel de Montaigne in France. We may, like Toulmin, also include William Shakespeare, although he employed a literary rather than a philosophical mode of inquiry. Montaigne is interesting also because he invented the essay genre in his famous work *Essais*, which literally translates as "attempts" from the French. Montaigne's essays begin from his own perspective and then wander into the world and around many different subjects from art and death to cannibalism. Nothing is left out, censored, or avoided by Montaigne, who wrote with honesty and curiosity about his own life and how it was related to more general topics in the 16th century in a way that prefigured the practices of qualitative researchers and contemporary autoethnographers.

With the impressionistic style of Montaigne and other humanists, we are far away from Plato's search for eternal and transcendent truths. The early modern Renaissance humanists were more interested in the real, messy, and complex features of human life than in developing ideal, abstract, and pure concepts. This changed, however, with the beginning of modern philosophy as such, often traced back to Rene Descartes, who was born in France in 1596 and who developed an impressive philosophical corpus until his death in 1650.

Modern Philosophy: From Descartes to Marx

So much has been written about Descartes and his influence, and he is very often misunderstood or outright demonized as having single-handedly invented all the dualisms that have haunted Western philosophy and science for four centuries now. This is unfair, for, as Robinson (2008) has pointed out, "Descartes is not the comic hero of 'Cartesianism' but a theorist strongly inclined toward philosophical materialism while, at the same time, having to acknowledge the conundrums arising from attempts to apply that materialism to the domain of consciousness" (p. 77). The background to Descartes' philosophy (now known as Cartesianism) was the emerging mechanical natural science during his time, represented for example by Galileo, who famously claimed that the book of nature is written in the language of mathematics. The natural world is a world of numbers, a quantifiable reality, which scientists should approach in mechanical terms, searching for what Aristotle had called the efficient causes of what happens. Against the whole Aristotelian conception of nature as a realm of meaning, value, and purpose, the new natural sciences viewed nature as devoid of meaning, purpose, and qualities. A stone does not fall to the ground because of anything teleological in the stone (e.g., because it comes from the ground and wants to return to it) but, rather, because of mechanical laws of nature. However, it is a stubborn fact that we experience not only quantities in our lives but also good and evil, beauty and ugliness, and many other qualities, so where do these qualities come from? The only answer seems to be that they come not from the world, but from the human mind— that is, from consciousness.

Here we have the problematic Cartesian distinction between *res extensa* (the outer, extended world) and *res cogitans* (the inner world of thoughts and experiences). For Descartes, what we experience are ideas in the mind. We cannot get out of our heads and get in touch with reality as it is, for he knew too much about the mediational role of the sensory system to be a direct realist (especially through the science of optics, which had demonstrated that outer objects are represented upside-down on the retina). So, for Descartes, we do not see the world, but only our representations of the world, which he called ideas. In a way, Descartes continued the idealism of Plato, but whereas Plato had placed ideas in the structure of cosmos itself, Descartes placed them in the minds of individuals, corresponding to what I referred to previously as a shift from a theocentric to an anthropocentric worldview. Ideas remained the building blocks of human knowledge, but Descartes transformed them from metaphysical entities and made them into mental ones instead. Four centuries later, the science of psychology would refer to these mental ideas as "mental representations," and the entire field of cognitive science would be based on their existence.

The well-known *Meditations on First Philosophy* by Descartes from 1641 employ a "method of systematic doubt" to arrive at what he argued was fundamental and indubitable knowledge. In this book, he takes the reader on a fascinating journey through various stages of doubt. It turns out that we cannot trust our senses (because we are familiar with illusions), and we might be dreaming or suffer from delusions created by an evil demon. Fundamentally, there is only one thing beyond all doubt, namely the fact that we can think and doubt, which leads Descartes to perhaps the most famous philosophical dictum of all time: *Cogito ergo sum*—I think, therefore I am. *This* is certain, according to Descartes. From that foundation, Descartes begins to build up our knowledge again, which also involves God (for since I have an idea of God as an infinite being, and I am finite myself, then that idea can only come from an infinite being, which demonstrates the existence of God). This epistemological process is akin to the process of educating military personnel that first breaks people down before building them up again in a new and improved version. And this is what Descartes is doing: Teaching his reader that systematic doubt can give us true knowledge—in the shape of what he called "clear and distinct ideas"—and it is a kind of knowledge that we each can grasp individually, as disembodied and nonsocial intellects—or what he called *res cogitans* in Latin. Essentially, as human beings, we *are* thinking things.

As such "thinking things," humans are detached not only from the world of extended things, which they know only through their representations of the outer world, but also from their own bodies. Thus, in Descartes' philosophy, we have a subject–object dualism and also a mind–body dualism, although he had a rather more sophisticated approach to this than most philosophers are aware of now. Robinson (2008) quotes with approval from Descartes' *The Passions of the Soul*: "We need to recognize that the soul is really joined to the whole body, and that we cannot properly say that it exists in one part of the body to the exclusion of others" (p. 76). However, elsewhere in Descartes' writings, we find the theory that body and soul interact in a part of the brain called the pineal gland (which looks like a small pine cone), and this has been much ridiculed by later philosophers because how can something physical and something mental (i.e., non-physical, according to Descartes) meet in a physical location in the body? This would seemingly have to presuppose that the non-physical entity was physical after all. Before laughing too loudly, though, we should bear in mind that even contemporary neuroscientists often commit similar logical mistakes when they address the relationship between mind and brain, analyzed, for example, by Bennett and Hacker (2003) in their magnum opus on the philosophical foundations of neuroscience.

In the years after Descartes, a string of British philosophers continued the work he had begun, although in a way that also involved significant adjustments. Descartes is known as a rationalist because he believed in the powers of the rational mind to arrive at true knowledge in relative independence of the senses. This was very different with Locke, Berkeley, and Hume—the so-called British empiricists—who wrote their treatises from the end of the 17th century and well into the 18th century. Locke, a medical doctor and philosopher, sometimes referred to as the father of liberalism, built on Descartes' ideational theory of knowledge, according to which we experience objects in the mind (ideas) rather than the things themselves. But unlike Descartes, Locke did not believe in innate ideas but, rather, argued that we obtain knowledge only through the senses. Initially, the mind is a *tabula rasa*, a blank slate, on which experience is written as the person receives impressions from the world. In his *Essay Concerning Human Understanding*, Locke operated with a significant distinction between primary and secondary qualities. Primary qualities are said to be properties of objects that are independent of an observer, and they include extension (cf. the Cartesian notion of *res extensa*), solidity, motion, and number. These are physical properties that would exist also in a universe without sentient creatures. Secondary properties are those that produce sensations in an observer, such as taste, smell, and color. We cannot obtain objective knowledge about these properties because they only really exist in the eye of the beholder. This is a familiar distinction for many qualitative researchers, who sometimes explain their inquiries with reference to this alleged subjective dimension as a supplement to the quantitative tradition that deals with properties that are independent of any individual observer.

Berkeley followed in the empiricist footsteps of Locke but argued that there is no philosophical warrant behind the distinction between primary and secondary qualities. If the latter necessarily arise and exist in the mind, so do the former, for we cannot leave the mind and arrive at an objective world outside. Berkeley formulated the principle of *esse est percipi*—to be is to be perceived. Things exist only insofar as they are perceived, and this is in a way a return to Plato's idealism, which posited that ultimately reality is made up of ideas. Berkeley's idealism, however, has a more subjectivist leaning because ideas exist in minds rather than the world (or actually, the world simply *is* the contents of minds). Here, he also followed Descartes' subjective idealism. But the question then quickly arises: What about those things that no one experiences? Like the tree that falls in the forest without anyone noticing it? Well, conveniently (we might say), Berkeley (who was a bishop) argued that such things are still perceived by God. Ultimately, the universe rests in the mind of God.

So, after the rationalism and nativism of Descartes, we see that Locke got rid of innate ideas and Berkeley annulled the mind-independent outside world, but there was still one radical step to be taken by David Hume, who can be said to take the empiricist philosophy to its end point: Hume argued that the self does not exist. The feeling, thinking, perceiving self is an illusion, according to Hume. He wrote his provocative *Treatise of Human Nature* in 1739, when he was just 28 years old, and the text still reads as a fresh and clear account that rivals today's deconstructionist writings in terms of radicality. If "to be is to be perceived," as Berkeley had argued, then we must say, according to Hume (1739/1978), that there is no self for the simple reason that this entity is not experienced:

> For my part, when I enter most intimately into what I call *myself*, I always stumble on some particular perception or other, of heat and cold, light or shade, love or hatred, pain or pleasure. I never can catch *myself* at any time without a perception. When my perceptions are remov'd for any time, as by sound sleep; so long am I insensible of *myself*, and may truly be said not to exist. (p. 252)

Hume's philosophy was the culmination of a huge movement in Western thought that had been on its way for centuries, referred to, by Kessen and Cahan (1986, p. 640), as "the great Western transcendental slide from God to Nature to Mind to Method." Hume's approach was based not on metaphysical assumptions about God, Nature, or even Mind but, rather, on Method, for Hume wanted "to introduce the experimental method of reasoning into moral subjects," which was the subtitle of his *Treatise* (Hume, 1739/1978, p. xi). Hume was greatly impressed by Newton's mechanical physics, which had made it possible to comprehend the physical world in terms of universal laws of nature, rendering it calculable and controllable (what follows builds on Brinkmann, 2011). It was in this mechanical, disenchanted world that Hume found himself, and it was here he set out to develop a Newtonian science of the mind. By introducing the experimental method of reasoning into moral subjects, Hume wanted to reform the science of man. He believed that all sciences, including mathematics and natural philosophy, depend on the science of man "since they lie under the cognizance of men, and are judged of by their powers and faculties" (Hume, 1739/1978, p. xv). Like his empiricist predecessors—Locke and Berkeley—Hume carried forth Descartes' representational epistemology according to which the mind, envisioned as a kind of container, is in direct contact with its given contents ("impressions" and "ideas" in Hume's words). Hume continued the Cartesian project of making epistemology the prime philosophical discipline, taking the lead from the following question: How can I, as a discrete, experiencing being, know anything about the external world? One of Hume's significant analyses concerned the idea of causality—that A (e.g., one billiard ball) causes B (e.g., another billiard ball to move). We can observe the two balls moving in turn, but the causality, Hume argued, can never be observed. We can only acknowledge that events of type A tend to be followed by events of type B in "constant conjunction," and this is really all there is to say about causality. In Chapter 3, I discuss how this theory was rehearsed by the positivists of the 20th century.

One thing we cannot know about—because there is literally nothing to know—is morality. For Hume, there are no facts about good and evil. Something is said to be good if it arouses positive feelings in the mind, and something is bad if one reacts negatively. Hume was thus a subjectivist concerning values because of his argument that no moral qualities can be found in the world (but only in the subjective minds of people). But he did note that different individuals tend to have quite similar values (something that science explained more than a century later with reference to Darwinian natural selection). Again, the qualities dealt with by ethics, aesthetics, and political philosophy are in the subjective eye of the beholder. Related to this is also the famous is–ought problem sometimes referred to as "Hume's law." Hume observed that an "ought" cannot be logically derived from an "is." Hume nowhere says exactly this, but in his *Treatise* he notes that in all the systems of morality he has seen, the author shifts suddenly from "is" and "is not" to "ought" and "ought not" (e.g., from "God *is* our Creator" to "we *ought* to obey him") without explaining this "new relation" (Hume, 1739/1978, pp. 469–470), and it is clear that Hume does not think that this can be justified. At the beginning of the 20th century, moral philosopher G. E. Moore introduced the term "the naturalistic fallacy" for the attempt to derive evaluations from natural matters of fact. Unlike Hume, Moore was no subjectivist, however. He did not think that moral values were subjective or nonexistent, but he argued that moral properties were *non-natural* and apprehended by intuition rather than science. Since Moore, "the naturalistic fallacy" has been used to name any attempt to derive evaluations from matters of fact, and this discussion is very often raised in relation to qualitative research: As a scientific enterprise, the defenders of the is–ought dichotomy argue, researchers can only describe *what is*. The question of *what ought to be* must be left to policymakers or voters or whomever may want to articulate their subjective values. This is also a significant discussion in qualitative inquiry: To what extent are qualitative research practices based on or related to values—can they ever be value-free, and, if not, then which values are worth subscribing to?

Although the term "psychology" was not used in English before 1748 (by David Hartley), Hume can be considered as the grandfather of psychology and the behavioral sciences because he wanted to develop a science of the mind, which was to be similar to Newton's science of the physical world. This was to be a science of relations between impressions and ideas and of how humans are motivated by their (irrational) passions. Our rational powers can only devise the means with which to realize our desires, but they can never dictate what we ought to desire. All this, however, was criticized on the continent by Immanuel Kant, the great German Enlightenment philosopher. Kant actually believed that a science of psychology was impossible exactly because the subject matter did not lend itself to quantification. And generally, Kant tried to refute the main points of Hume's philosophy, which famously is said to have provoked his (Kant's) awakening from a "dogmatic slumber." Unlike Locke, Berkeley, and Hume, Kant was no empiricist, but nor was he a pure rationalist like Descartes. In a way, his philosophy united these two strands of thought by insisting that the categories of the mind and the contents of experience are equally important in the process of knowing. His philosophy is transcendental because it begins from the fact that we do have experience and then works from the question, How is this possible? A transcendental argument is about the conditions of possibility for something that exists. Kant's starting point was this: Experience exists—how is it possible? Concerning Hume's views of causality and natural science, Kant thought it was scandalous to refer these to the subjective realm of the individual mind. Instead, he argued, there are certain rational prerequisites to experience that we can never get rid of, and without which there simply could not be experience. These are space and time and also categories such as quantity, quality, relation, and modality. Kant's philosophy is quite technical, but the key point is that experience is constructed when sensory contents meet the necessary and universal forms of rationality. For that reason, Kant is often referred to as a constructivist, and this also applies to his analysis of morality.

Kant also believed that Hume's view of morality was scandalous because it made this otherwise worthy human phenomenon wholly contingent—that is, relative to whichever passions human beings might have. Against this, Kant argued that morality is necessary because it springs from a universal practical reason, which states—in the well-known categorical imperative—that an action is moral if, and only if, the agent can will that its maxim (or principle) should become a universal law of nature. Thus, if one can rationally will that the principle behind an action (e.g., "help people in need") should always apply—as if it happened by itself—then we can say that it is a moral action because it is universalizable. Therefore, morality has nothing to do with emotions (like Hume's passions) or the subjective viewpoint, but springs from universal reason as such. There is much more to say about Kant, of course, but for qualitative researchers it is interesting how his constructivism prefigured many later schools of constructivism—that is, the theory that knowledge is a mental construction. Kant's version of constructivism, however, was more universal than most current varieties. For Kant, our knowledge is constructed when the sensory contents meet the apparatus of the mind, but the operations of the mind are universal because humans are rational creatures—and rationality is something universal in the Kantian scheme.

The abstract universality of Kant's philosophy was already criticized by G. W. F. Hegel when the 18th century became the 19th century. Hegel can be thought of as the grandfather of sociocultural studies because he relativized Kant's universal categories of reason and tied them to specific times and places. What Kant called *Moralität* (universal and abstract morality), for example, was seen by Hegel as related to the *Sittlichkeit* or the ethical life as actually and concretely lived. There was still a sense of necessity in Hegel's thought, however, but this manifested itself in how the ideas on which societies are built transform themselves through time—rather than in anything that transcend time and place. Hegel detected an order in history related to what he called the self-realization of spirit (Hegel, 1807/1977). This Hegelian idea is quite difficult to grasp, but it means that there is an order in history related to a gradual development toward increased self-reflexivity. As history progresses, humans become capable of relating to themselves in new ways because societies become freer and more transparent. For Hegel, there is thus progress in the form of an unfolding of ideas through history.

This perspective was continued in a materialist direction by Karl Marx, who looked at the structure of the means of production in a society as the engine of history—rather than human ideas and ideals. Philosophers like to say that Marx turned Hegel on his head because Marx gave priority to the material dimension of society—with all its technologies, organization of labor, and class divisions—rather than to the ideas that are formulated in the course of history. We can say that Marx gave priority to practice over theory, whereas Hegel emphasized the primacy of theory (ideas) over practice. On this point, the ancient discussion between Plato and Aristotle once again played out—this time in the 19th century between Hegel and Marx. In his 8th thesis on Feuerbach, Marx (1888) said that "all social life is essentially practical. All mysteries which lead theory to mysticism find their rational solution in human practice and in the comprehension of this practice." And Marx not only wanted us to stand at a distance and analyze the practices of society. Rather, he wanted us to change them. The famous 11th and final thesis on Feuerbach reads as follows: "The philosophers have only interpreted the world, in various ways; the point is to change it." This activist impetus spoke to numerous Marxist scholars in the 20th century; it spoke to critical theorists such as Adorno, Horkheimer, Habermas, and Honneth, and it certainly speaks to many qualitative researchers more than 100 years later.

Not just Marx but also more idiosyncratic figures such as Kierkegaard and Nietzsche reacted to Hegel's systematic idealism. Kierkegaard developed an existential philosophy, which focused on the basic existential conditions of life related to freedom, anxiety, and despair. Kierkegaard reacted to the historicist straitjacket of Hegel and proclaimed that truth is related to subjectivity. This was continued in different ways in the philosophies of Heidegger and Sartre in the 20th century. Nietzsche found inspiration in Greek art and tragedies and Darwinian evolution and sought to trace the development of human ideas relative to a will to power. This genealogical method (looking at the real historical contexts of ideas) was continued by Foucault in the 20th century, as he went on to study how power relations affect human knowledge (this is discussed in-depth in Chapter 6). With Marx, Kierkegaard, and Nietzsche, we are nearly ready to begin the next chapters, which examine different streams of thought that have been more directly relevant for the development of qualitative research practices from approximately the beginning of the 20th century.

Summary and Looking Ahead

Before moving on, it is time to sum up and ask what we have learned about the historical development in Western ideas from the Greeks more than 2000 years ago and philosophers such as Kant, Hegel, and Marx in the 18th and 19th centuries. Perhaps the most general conclusion is that philosophy's original questions about what there is—about the nature of the world—were gradually replaced by the more anthropocentric questions about what a human being is and how the human being can know anything. The metaphysical approach of Plato, which had placed ideas in the world as such, was gradually replaced by the modern epistemological approach of Descartes and his followers, which placed ideas in the mind itself. Epistemology took over as the philosophical starting point in line with the historical development of individualism. A Marxist analysis of this very transformation would view it as following from fundamental changes in the modes of production. With the uprooting of previous premodern societies and communities and the establishment of modern nation states with industrial production, the individual entered world history as a key element. The emergence of mechanical natural science with

Galileo and later Newton was also very important as a mediator of this development, because it led to a breakdown of the teleological universe of the Greeks (where meaning, value, and purpose were part of cosmos), and paved the way for the modern scientific understanding of human beings—for example, in the early psychology of David Hume, which depicted man as driven by passion in a value-neutral world.

The very idea of qualitative research is a child of modernity's split—represented most clearly by Descartes, Locke, and Hume—between the objective and the subjective, and between *quanta* and *qualia*. This split became significant with the birth of modern natural science and gave rise to the question of how to study those aspects of the world that do not seem to fit the perspective of the physical sciences. As discussed in the following chapters, this question was answered in different ways by thinkers in the United Kingdom, Germany, the United States, and France. What we are left with so far is both the epistemological problem (How can a subject detached from the world know anything about this world?) and the problem *of* epistemology (Is it really helpful to frame the problem in terms of modern epistemology?). We are also left with an is–ought dichotomy related to the dualisms of mind and body, subject and object, individual and society, and theory and practice. I previously argued that the whole epistemological problematics comes from what can be called an "epistemology of the eye," which in many ways runs like a thread from Plato to contemporary philosophers (Brinkmann & Tanggaard, 2010; I discuss this further in Chapter 5).

Since the Greeks, as we have seen, the notion that ideas are "out there" has been fundamental. For Plato, ideas are "out there" as the basic, unchanging constituents of being (the Platonic "Forms") that we humans may come to recognize because we are endowed with immortal souls that stem from the same realm of ideas. Learning something means for Plato a "turning of the souls" away from mere phenomena so that humans may come face to face with the eternal ideas. Plato's guiding imagery thus draws on light and visual metaphors of knowledge. In the allegory of the cave, the sun is likened to the overarching idea of the good as that which brings light to all other ideas so that they may be seen. Knowing is seeing. Learning happens through visual confrontation with something. And the mind—the soul—is that which sees, a "mirror of nature" in Richard Rorty's (1980) illuminating (notice again the light metaphor) words. And although Aristotle transformed much of Plato's philosophy into a more viable, scientific approach, the visual metaphors lived on, for example, in his "hylomorphic account of knowing" (p. 35), according to which reality impinges on our senses, just as wax can receive an impression of a signet-ring. With the subsequent ideational and representational epistemologies of Descartes and the British empiricists (Locke, Berkeley, and Hume), ideas are finally transformed from outer cosmic constituents to inner mental entities that humans build up "in their minds" in order to know the world. In recent decades, modern cognitive science has continued the project of charting how ideas (mental representations) copy the world. Thus, a whole epistemology of the eye has been at work in Western thought. This has not just influenced our theories of knowledge, truth, learning, and the mind but has also had enormous practical implications, not least in educational contexts. It has also been influential in the ways we think of research in general and qualitative research in particular, and some of the developments in the 20th century—such as phenomenology, pragmatism, and the discursive approaches following from Wittgenstein's groundbreaking philosophy—tried in various ways to question and overcome the epistemology of the eye.

In the following chapters, I present four different philosophical approaches to qualitative research that have been important historically and continue to be so. These approaches have four different birthplaces in Great Britain (positivism and realism), Germany (hermeneutics and phenomenology), the United States (the pragmatisms), and France (structuralism and post-structuralism). They have all emerged from the history of philosophy that I have recounted in this chapter. For each geographical birthplace, I first provide some background about the philosophical ideas and then move on to describe how these ideas have been translated into methods and techniques in qualitative research.

3

BRITISH PHILOSOPHIES OF QUALITATIVE RESEARCH

POSITIVISM AND REALISM

IN A MODERN SOCIETY, influenced by and even dependent on scientific knowledge, it seems intuitively important that the work of researchers should be as reliable and objective as possible. Political decisions are often made with reference to scientific studies, for example, so these should be transparent and trustworthy. For many people, this also means that the results of researchers' studies should not be tied to the subjectivity of the individual researchers. If something is true, it is not so simply because John Smith says it is true—or because of his authority or background—but because of certain facts in the world that can be discovered by the use of scientific methods. These methods ideally enable others to conduct a study similar to John Smith's and to assess the objectivity (and the validity and reliability) of what John Smith has said.

The previous paragraph probably represents many people's spontaneous understanding of scientific practice. At least since the Enlightenment, philosophers and scientists in the Western world have developed this idea of science, and it has become common sense today. It is this kind of approach to scientific work that positivism has tried to purify and translate into methodological procedures. It turns out, however, that attempts at doing so quickly run into problems, as discussed later. Also, critical realists, who are partly very critical of positivism, have sought to articulate an approach to qualitative research along the scientific lines that many people intuitively share: The goal of scientific practice is to go beyond what is immediately present, including common opinions and prejudices, in order to arrive at a truer understanding of reality. This is what Noblit and Hare (1988) call "making the hidden obvious."

Positivism

The term *positivism* figures in most—if not all—qualitative research textbooks that I have ever read. However, it is very rare that the authors of these textbooks provide concrete references to positivist philosophers, and the term itself usually appears as a simple term of abuse, signifying "what we are against" as qualitative researchers. *They*—the positivists—are objectivist, realist, and solely employ quantitative methods, whereas *we*—the qualitative researchers—give voice to individual experience, are anti-realist, and are ethically superior. This is misguided in more than one way, as discussed later. First, it is simplistic—and in several cases simply wrong—to claim that positivist philosophy cannot incorporate qualitative methods. Second, it is problematic to claim that positivism is a form of realism. In fact, philosophers of science often characterize it in the exact opposite way—as a form of anti-realism. One can even say that realism as a philosophy of science explicitly emerged, in the British tradition, out of a critique of positivism (Harré & Madden, 1975), namely in an attempt to say that positivism misses *the real* that is hidden behind the world that we experience. But let us now focus directly on this *bête noire* of positivism.

Positivism could have been placed in the track of French ideas because its founder, Auguste Comte (1798–1857), was French. It could also have been placed in the German track because its most influential 20th-century representatives—the Vienna circle positivists—spoke German. But it nonetheless makes sense to place it in the British tradition because its great grandfather was British, or more precisely Scottish, and bears the name David Hume. I introduced Hume in Chapter 2 as one of the most important modern philosophers to come after Descartes. Hume was an iconoclast who generally proceeded philosophically with Occam's razor in his hand. As already mentioned, Hume saw no need for

God, the self, or an idea of causality beyond human experience. And it is this third point that makes him the great grandfather of positivism. For positivism is distinguished among the philosophies of science by its insistence that the only legitimate knowledge claims are those that refer directly to experience—that is, to what can be positively verified. And just as we do not experience a self in our various experiences, according to Hume, so we do not experience causality. We experience events of type B regularly happening after events of type A, and that is all we can really mean by saying that A is the cause of B, according to the positivists. This, in a nutshell, is positivism. Scientific practice should consequently proceed by using systematic methods to study the constant conjunctions in our experience between different kinds of events (e.g., A's and B's). This applies to the human and social sciences just as much as to the natural sciences.

But let us take a step back and ask where the term positivism itself comes from. It was created by Comte, the first positivist proper, who founded not only positivist philosophy but also the science of sociology. His positivist philosophy reacted against religious dogma and metaphysical speculation and advocated a return to what is positively observable—the data—and he famously argued that human knowledge passes through three different stages: the theological, the metaphysical, and the scientific or positive stage (Comte, 1830/1988, p. 1). In the first theological stage, "the human mind directs its researches mainly toward the inner nature of beings" (p. 2) rather than "abstract forces" (the metaphysical stage) or "the connection established between different particular phenomena and some general facts" (the positive stage) (p. 2). The movement away from what Comte calls theology represents the transition from an Aristotelian worldview, where things are thought to move according to their inner natures, through a metaphysical approach that postulates transcendent forces beyond human experience, to finally arrive at positivism and its emphasis on direct observation of connections between various events in experience.

Comte argued for the need of social engineering on a scientific basis to establish social order under the beginning anomic conditions of modernity. Once we know about the laws that govern social life, we can begin to make use of our knowledge to make society better. This meant that he had to ask about the direction in which society was to move; he had to inquire into values and human morality, for what does improvement actually mean? Comte begun, but never finished, a grand work on the science of *la morale*—concerning values and morality—but it was certainly not in the burgeoning science of psychology that he would look for answers. It was Comte's general conviction that there are two classes of phenomena of interest to human scientists: biological/organic, on the one hand, and societal/collective, on the other hand. The first class of phenomena was to be dealt with by biology, whereas the second class was the province of "social physics" or "sociology" (the word itself was Comte's invention)—and this left nothing for psychology to study (Cahan & White, 1992). There is, according to Comte, no ontological realm of individual mental phenomena to be studied by psychology between the biological and the sociological sciences. In that way, he was as starkly antisubjectivist as many of today's post-structuralists (see Chapter 6), who have argued that the philosophy of the subject, humanism, or the philosophy of consciousness is dead.

Comte never wrote his treatise on the science of *la morale*, but we know that he called this science-to-be "a sacred science" (Samelson, 1974, p. 221) and found that "science proper is as preliminary as are theology or metaphysics, and must finally be eliminated by the universal religion" (p. 222). The science of *la morale* was to culminate in a form of communitarian, human religion, emphasizing "the dependence of man on society (humanity in its historical totality), [and it] considered the isolated individual a false abstraction" (p. 222). It is quite clear that Comte's *morale*, like his positivism in general, contained much more than pure methodology, and he did not take the stance of a value-free or politically neutral scientist but, rather, resembled Marx in his insistence on the importance of the historical situatedness of human behavior. I mention this in order to underscore the richness, even occasional intellectual wildness, of much of the theorizing that went on among the original positivists, especially Comte.

The influence of positivist sociology can be seen in the later work of Emile Durkheim, an early sociologist who gave penetrating qualitative analyses of social phenomena (the following reworks passages from Brinkmann & Kvale, 2015). And it is also clear that positivism had an extended influence on the arts of the 19th century, inspiring a move from mythological and aristocratic themes to a new naturalism, depicting in detail the lives of workers and the bourgeoisie (for some of this history, particularly in the British context, see Dale, 1989). In histories of music, Bizet's opera *Carmen*, featuring the lives of cigarette smugglers and toreadors, has been depicted as inspired by positivism, and Flaubert's realistic descriptions of the life of his heroine in *Madame Bovary* warrant a characterization as a positivist novel. Impressionist paintings, sticking to the immediate sense impressions, particularly the sense data of pointillism, also drew inspiration from positivism. Michel Houellebecq is a contemporary French author who explicitly acknowledges his inspiration from Comte's positivism, and Houellebecq has written the preface for a recent volume on Comte today (Bourdeau, Braunstein, & Petit, 2003). The early positivism was also a political inspiration for feminism, and it was the feminist Harriet Martineau who translated Comte's *Positive Philosophy* into English. In philosophy, the founder of phenomenological philosophy, Husserl, stated that if positivism means being faithful to the phenomena, then we, the phenomenologists, are the true positivists. It can even be argued that the insistence in Comte's positivism to stay close to observed phenomena rather than engaging in metaphysical speculation about theoretical entities comes close to a postmodern emphasis on the importance of staying close to observable surface phenomena rather than postulated deep structures. For positivists as for postmodernists, the surface has become the essence (see Chapter 6).

I have tried to present the birth of positivism with Comte as a much more vibrant and many-sided philosophical project than what is usually thought of when the term positivism is mentioned. After Comte, however, positivism traveled back to Great Britain, especially with the philosophy of John Stuart Mill. Mill was a very influential philosopher, both in ethics and in political philosophy in which he developed a utilitarian position, but also in the philosophy of science. He was the first to coin the term "moral science," which became very influential in 19th-century discussions about the human sciences. *The Logic of the Moral Sciences* was originally part of his magnum opus, *A System of Logic*, from 1843 (Mill, 1843/1987), but it was subsequently published separately. It was translated into German in 1849 by Schiel, and the word he chose was "*Geisteswissenschaften*" (Kessen & Cahan, 1986, p. 649), which became hugely important for the subsequent hermeneutic thinkers.

Some scholars also consider Mill to be responsible for having created a modern scientific psychology (Gazzaniga & Heatherton, 2003; Miller, 2004), significantly inspired by Hume's Newtonian and mechanical science of the mind. Mill argued that the backward state of "the moral sciences" could be remedied only by applying to them the methods of physical science (Mill, 1843/1987, p. 19), which was in direct continuation of Hume's imperative to introduce the experimental method of reasoning into moral subjects. Unlike Comte, Mill thought that there in fact could be such a thing as scientific psychology. Its subject matter would be "the uniformities of succession, the laws, whether ultimate or derivative, according to which one mental state succeeds another—is caused by, or at least is caused to follow, another" (p. 38). This is a reiteration of Hume's approach, according to which psychologists and other human scientists should study the causal chains of experienced phenomena.

The open approach of Comte's classical positivism, and to some extent that of Mill, was gradually lost in the methodological positivism of the Vienna Circle in the 1920s. Philosophers from the Vienna Circle, such as Schlick, Carnap, and Neurath, are probably the names that many contemporary scholars think of in relation to the term positivism. Neurath's (1929/2003) statements in *The Scientific World Conception* are illustrative: The goal of scientific practice is to arrive at "unified science," which means to "harmonise the achievements of individual investigators in their various fields of science" (p. 31). Thus, ultimately, what goes on in physics, chemistry, biology, psychology, sociology, and so forth can (and should) all be put coherently together, and some "physicalists" would even claim that all knowledge will in the long term be reducible to physics and be expressed in its language. Other remarks from Neurath are interesting—for example, when he writes about positivism:

> Neatness and clarity are strived for, and dark distances and unfathomable depths rejected. In science there are no "depths"; there is surface everywhere: All experience forms a complex network, which cannot always be surveyed and can often be grasped only in parts. Everything is accessible to man; and man is the measure of all things. (p. 31)

The last sentence is a reference to the pre-Socratic sophist, Protagoras, who claimed that man is the measure of all things.

The almost poetic valorization of surface rather than depth is also close to postmodernism and the Wittgensteinian insistence that "nothing is hidden." For us positivists, Neurath continues, "something is 'real' through being incorporated into the total structure of experience" (p. 33). Although coming from a very different paradigmatic background, this is not far from the broad phenomenological outlook of many qualitative researchers, who focus on "lived experience." For a positivist such as Neurath, knowledge must be grounded in the experience of human beings and not in anything metaphysical or transcendent behind or beyond our experiential realm.

Sometimes this is formulated by linguistically oriented positivists as the principle of verification: A sentence is meaningful insofar as it can be verified with reference to human experience. According to positivists, this excludes metaphysical and theological sentences from being meaningful. A. J. Ayer, the main protagonist of positivism in the United Kingdom, wrote the classic *Language, Truth and Logic* (Ayer, 1936/1990),"which founded logical positivism—and modern British philosophy," according to its book cover. It simply begins with a chapter brutally titled "The Elimination of Metaphysics." Ayer expressed the fundamental principle of verificationism very clearly: "We say that a sentence is factually significant to any given person, if, and only if, he knows how to verify the proposition which it purports to express" (p. 16). Thus, only if one knows which observations that will lead one to accept a proposition as true does one know what the proposition means. The sentence "the cat is on the mat" is meaningful because one may undertake certain operations in order to identify a cat, a mat, and the structural relation between them (one being "on" the other). Other sentences, such as "God is a trinity" or "John is fixated in an Oedipal conflict," are not similarly meaningful because it is not (Ayer argues) possible to verify them experientially. In addition to statements that can be verified in experience, positivists also believe that logical statements, or so-called analytical statements, are meaningful. These are sentences that are logically derived from other sentences. Strictly speaking, these are tautological, which means that they do not teach us anything substantially new (e.g., from the sentence "the cat is on the mat," we may infer that "a mammal is on the mat" because a cat is by definition a mammal). As Neurath (1929/2003) summed up, "The scientific world-conception knows only empirical statements about things of all kinds, and analytic statements of logic and mathematics" (p. 33).

Needless to say, this provides for a very narrow approach to science and meaningful statements, excluding not only theological and metaphysical sentences but also ethical and aesthetic ones. And, even more problematic for the positivists, it seems to exclude their own philosophical statements, including the verification criterion itself, from the realm of meaningful sentences. This kind of critique—that is, an immanent critique based on the assumptions of what one is criticizing—has often been raised against positivism. "If *this* is how you define meaningful statements," one can say to the positivist, "then your own definition is meaningless, because it cannot be verified in experience!" Or, in relation to the positivists' exclusion of values from rational discussion and scientific practice, one can say, "If you define science as value-free, because you believe that it should strive for objectivity and reliability, then you forget that objectivity and reliability are *also* values!" (this, in effect, is Putnam's (2002) critique of the fact/value dichotomy often invoked by positivists).

In my account of positivism so far, I have tried to show that it has emerged historically in at least two or three waves. First, in the early and rather "wild" form with Auguste Comte. Here, there was little or no opposition to what we would today refer to as qualitative research. Comte believed that social phenomena were too complex to be subjected to mathematical analysis, and the application of mathematical analysis was in any case not necessary for a positive science: "Our business is to study phenomena, in the characters and relations in which they present themselves to us," Comte said, "abstaining from introducing considerations of quantities, and mathematical laws, which is beyond our power to apply" (quoted in Michell, 2003, p. 13). What Joel Michell has called "the quantitative imperative" in psychology and related sciences does not come from positivism (not even from the later versions of Carnap or Neurath) but is connected to a bureaucratic culture that is obsessed with quantitative evidence and a whole "what works" ideology (Kvale, 2008). Thus, my argument is that despite problematic features of positivist philosophy (some of which are addressed by the later realist philosophers), qualitative researchers should be careful not simply to dismiss a straw man when rejecting positivism out of hand. A careful study of especially Comte will reveal a much more open and complex philosophy than what we conventionally think of in relation to positivism. The inductive approach of positivism, which strives for a gradual creation of human knowledge through increasingly more experience, is actually in line with much qualitative research. And of course the focus on experience itself—what is sometimes called *phenomenalism* because of the belief in immediate phenomenal experience as foundational for human knowledge—is also close to qualitative researchers' respect for experience.

Positivism has today been transformed from a philosophy of science into a more bureaucratic procedure of producing and applying knowledge. The whole movement toward "evidence-based knowledge" and the "what works" agenda in the human and social sciences—finding inspiration in the Cochrane movement in medicine from the 1970s onward—can be considered a kind of "bureaucratic positivism," which is probably a much greater threat to qualitative research than the original positivism of Comte and also the second-wave positivism of the Vienna Circle. For the politics of evidence that is part of bureaucratic positivism often sidelines and excludes qualitative studies of "what there is" in the world in favor of randomized controlled trials of "what works" according to quantitative measures, and sometimes even claims that only the latter kind of study is scientific.

Realism

After positivism followed a broad and exciting range of responses in the world of philosophy of science. The theory of science developed by Karl Popper from the mid-1930s onward accepted much of the positivist agenda, albeit with some significant adjustments (Popper, 1959). Popper famously argued that scientific truth cannot be asserted through verification. The kind of verificationism posited by the positivists could not lead to genuine knowledge, Popper asserted, because of the problem of induction: No matter how many instances of something that we positively observe, we will never be in a position to arrive at general knowledge. Often, philosophers use the example of the white swans to explain this. Even if all observed swans have been white, we can never know if all swans are actually white (and in fact they are not) and will remain white in the future. So, if science proceeds in accordance with positivist principles through verification and inductive inference, it means that no secure and general knowledge exists. Popper's solution to this problem was to insist that scientific practice does not in fact work inductively but, rather, deductively. Scientists formulate bold conjectures of a general kind, which are then subjected to empirical investigation. And those conjectures that survive critical study are maintained until further notice. Those conjectures that are falsified should be abandoned. Thus, Popper replaced verification with falsification as the mark of scientificity. A statement or theory is not scientific because it can be verified in experience but, rather, because we may (in principle) falsify it. Popper believed that Marxist and Freudian theories were unscientific because they could never be falsified. If the Freudian analyst presents an interpretation to the patient, for example, and the patient agrees, then this is taken as verification of the interpretation. But if the patient disagrees, then this is taken as a kind of denial, which also verifies the interpretation. At least this is the caricature of Freudian psychoanalysis that Popper had in mind, which guards it from being falsified.

Unlike the positivists, Popper was a scientific realist. His falsificationist theory of science led him to deny that scientific theories can be absolutely true, but there is no doubt that he believed that the objects of science have an existence beyond human theories and representations, and this makes him a scientific realist (Leplin, 2007). This is also what aligns him with the critical stance of "making the hidden obvious," and Popper himself referred to his philosophical position as critical rationalism. Positivism and realism are often confused today. But if realism is the view that there are working mechanisms in the world that cause more superficial aspects to appear and become observable, then most positivists were quite starkly anti-realists. They argued that we should stick to what can be positively observed and verified in experience (sometimes referred to as phenomenalism), which led them to distrust theories about underlying mechanisms or transcendent metaphysical structures. Causation on the positivist interpretation, as we have seen, simply meant constant conjunction in experience. For Popper and the later realists, this was not satisfactory because it made the scientific enterprise too relative and dependent on human experience.

Before moving on, we should pause and consider some of the other notable critiques of positivism that emerged in the mid-20th century. One of these was based on the idea that all observations are theory-laden. This was already voiced by Popper, but it is more often associated with the likes of Hanson, Quine, and Kuhn. In *Perception and Discovery*, Hanson (1969) referred to basic perceptual phenomena to argue that what we see in our scientific practices is not just "what is there" (as elementary positivist "sense data," for example) but also results from our theoretical interpretations. "There is more to seeing than meets the eye" is a slogan from Hanson's work. According to Hanson, it is not possible to draw a firm line between what we see and what we believe about what we see, for the latter influences the former. Before this, Quine (1951) had already argued against what he called the "dogmas of empiricism"—that is, the central tenets of positivist philosophy—that it is not possible to make a clear distinction between so-called analytic and synthetic statements. The former are statements that are logically true, whereas the latter are true because they correspond to matters of fact. Again, Quine's argument went against the positivists' neat separation of different kinds of statement, and Quine introduced the idea that the same set of empirical observations may result in two equally coherent theories that are nonetheless logically incompatible. Thus, we simply cannot move in a simple way from empirical observations to theory-building. Scientific data "underdetermine" scientific theories, which means that our observations are infused with theory and vice versa. Finally, in *The Structure of Scientific Revolutions*, Kuhn (1970) developed the notion of scientific paradigms as central to scientific practice. Paradigms are whole sets of theories and practices that enable something to emerge as an observation or a scientific fact. Thus, what counts as a fact or an observation is relative to the paradigmatic context in which it occurs, and the positivist idea of elementary sense data is thereby discarded. Notably, Kuhn referred to paradigms as entire "worldviews" and argued quite controversially that scientists live in a different world after a paradigm shift. This is so because what counts as the essential properties of the universe that can be observed may change after a paradigm shift. The Aristotelian natural philosopher saw teleological movement when looking at the stones and stars, whereas the post-Newtonian scientist saw mechanical movement effectuated by gravitation. Many subsequent philosophers of science have interpreted Kuhn as a kind of anti-realist social constructionist, but it is quite clear that Kuhn himself wanted to defend a version of realism (see Niiniluoto, 1996). According to Kuhn, there is a world independent of human sayings, doings, and interpretations, but this world shows itself in different ways relative to how we approach it with our various paradigms. Thus, Kuhn may be said to combine an epistemological relativism or perspectivism with ontological realism—quite like the critical realists that I return to later.

The post-positivist philosophers of science—such as Popper, Hanson, Quine, and Kuhn—are mainly relevant for qualitative researchers because of their critiques of positivism. They were mostly engaged with understanding the natural sciences. But there is a more explicitly realist line of thinking in philosophy that is more immediately relevant for human and social scientists. This line of thinking adheres to the dictum of Karl Marx mentioned previously that "all science would be superfluous if the outward appearance and the essence of things directly coincided" (Marx, 1894). For Marx, science is needed to go beyond appearance or the surface to critically uncover a truer reality. This goes directly against positivism, which was very skeptical about underlying essences of things.

The most common realist philosophy of social science in recent years has been *critical realism*, originally associated with Roy Bhaskar (1975/2008). Critical realism in Bhaskar's version is a dialectical theory, which builds on both Hegel and Marx, and which is concerned with not only epistemology but also ontology proper. Bhaskar saw positivism as nothing but an updated version of Hume's empiricism, and he also read Popper's critique of positivism as operating within the same set of fundamental assumptions that Popper had himself criticized. The starting point for critical realism, therefore, is a critique of the tendency to put human experience in the center of scientific knowledge, as represented by both the positivists and the phenomenologists (see Chapter 4). Instead, science should aim to study those structures and mechanisms that allegedly *generate* the phenomena that we experience. Bhaskar thus outlined an ontology that involved three levels or domains in the world: the *empirical* (our observations and experiences), the *actual* (events and phenomena), and the *real* (structures and mechanisms that operate causally).

In the world of human culture and society, the deepest domain of the real consists of generative mechanisms that are independent of minds and concrete societies. These mechanisms do therefore not change relative to how we choose to describe them—for example, in our scientific theories. It is thus possible to obtain genuine theoretical knowledge about them (cf. Aristotle's categories of knowledge depicted in Table 1.1). Thus, what Hacking (1995) has called "the looping effect of human kinds"—the interaction between how we describe humans and their actions, on the one hand, and the humans that are described, on the other hand—is rejected. There is a domain of the social world that is independent of how it is described and analyzed. The social world is not experiential or interpretative "all the way down," for, ultimately, we can reach a level of causal mechanisms that simply are what they are, and the most important goal of social science is to formulate an explanation of these. Unlike positivists, who (like Hume) reduced causality to "constant conjunction" in experience, and also unlike phenomenologists and interpretivists, who reject the idea that the social world is to be explained causally in the first place (see Chapter 4), critical realists maintain that qualitative social science needs to search for causal mechanisms when studying the world. What this means might become clearer in the next sections, which discuss different ways in which the British traditions of positivism and realism have influenced qualitative research.

Positivism in Qualitative Research

Although very few, if any, qualitative researchers self-identify as "positivists" today, much standard qualitative work can be labeled "qualitative positivism." As I have tried to show previously with particular reference to Comte, this is not a contradiction in terms because there need not be an opposition between qualitative inquiry and positivism. Quite the contrary in fact, given the experiential and phenomenalist focus among many positivists. If we understand positivism as the idea that researchers should stick to what it is possible to positively verify in experience, and with the related idea that specific scientific methods are what enable researchers to develop reliable and valid knowledge about various phenomena, then a positivist approach is extremely widespread among qualitative researchers.

A key aspect of positivism is that methodological procedures rather than subjective judgment or theoretical frameworks are said to determine scientific truth. In qualitative work, there are some varieties of grounded theory (but certainly not all of them) that come close to this. The ambition of grounded theory is to build knowledge through induction by using methods of coding and categorization of one's empirical data (Glaser & Strauss, 1967). Especially the early formulation of grounded theory, as led by Barney Glaser, has been interpreted as involving a rather strong positivist outlook (see Bryant, 2003). There is quite clearly an emphasis on how to "collect data" and analyze them through "coding" and "categorization," with the ideal being a process that can be relatively independent of the persons doing the work.

In many contemporary applications of CAQDAS—that is, computer-assisted qualitative data analysis software—often developed along the line of grounded theory methodology, there are inbuilt presuppositions that resemble positivism. The idea that data are in principle separable from the contexts in which they were produced, from the persons producing them, and from theoretical frameworks that are used to make sense of them seems to apply to many uses of CAQDAS. Increasingly, "data collection" is conducted in research teams, and when the data are uploaded to the computer system, they can be coded and analyzed by many different people. This kind of objectification and depersonalization of data (for lack of a better word) has emerged out of a positivist approach to research and is very different from phenomenological and hermeneutic approaches (as will be discussed in Chapter 4), in which analysis and understanding are related to a "fusion of horizons" between different people, which means that personal and contextual features become much more prominent.

Realism in Qualitative Research

The realist critique of positivism and the critical realist alternative has been well articulated by Maxwell (2012) in his idea of a realist approach to qualitative research. Unlike the anti-realism inherent in the phenomenalism of positivism, Maxwell's critical realism combines a realist ontology (the world contains entities and properties that exist independently of us) with a constructivist epistemology (our knowledge is constructed through human activities). In this sense, critical realism is a very general philosophy, but when applied to qualitative research specifically, the central argument is that we need to accept a causal role for meanings (p. 20). Meanings are not just contingent and socially constructed interpretations but also a real force in social life that can operate beyond human awareness. As we have seen, the positivists denied an ontological status to causes, but the realists affirm the concept of cause as referring to the inner workings of some process or phenomenon (p. 37).

For positivists, data are whatever we collect from experience and analyze, but for (critical) realists, data are understood as "*evidence* for real phenomena and processes (including mental phenomena and processes) that are not available for direct observation" (Maxwell, 2012, p. 103). We thus see very clearly how scientificity for realists is connected to "going beyond" what is immediately observable, or—to use Noblit and Hare's (1988) terminology—making the hidden obvious. The methods for doing this are conventional qualitative methods, according to Maxwell: Participant observation, narrative inquiry, intervention, and comparative studies are mentioned as methods for causal inference (p. 43). Almost anything goes methodologically, as long as the chosen method directs the researcher to the generative working mechanisms of the social world. Maxwell thus argues for what he calls validity based on an understanding of the phenomena rather than validity based on specific research procedures as in positivism (p. 144). Realists are not content to discover *that* some A causes some B to happen but, rather, argue that the scientific task is to explain *how* A causes B to happen. Maxwell highlights the key "realist principle" that "mechanism + context = outcome" (p. 164). It is not enough to describe and understand what happens in contextually framed situations. For the realist, the goal is to go beyond the situation and pinpoint the mechanism that causes something to happen, given specific contextual features. A classic example of this is unfolded in Box 3.1.

Summary

In this chapter, I have depicted a development in philosophical ideas from positivism to realism, including its so-called critical variety. Positivism is often misunderstood by qualitative researchers and

Box 3.1 **Realism and How to Make the Hidden Obvious**

Critical theory is a broad approach in the social sciences originating with Horkheimer and Adorno and followed by Habermas, Honneth, and other so-called Frankfurt school scholars. Critical theory combines empirical social science with philosophical reflection and is in line with critical realism as outlined here. Arguably the most famous qualitative research project conducted by critical theorists is the study of the authoritarian personality by Adorno and co-workers (Adorno, Frenkel-Brunswik, Levinson, & Sanford, 1950). In the wake of World War II, the researchers were interested in understanding the rise of anti-Semitism and fascist ideology and why it was so widely accepted in many European countries. The study, which employed qualitative psychoanalytic interviews, used a sophisticated interplay of open qualitative interviews and highly structured questionnaires for producing and validating data (see the description in Brinkmann & Kvale, 2015). An important part of the investigation consisted of psychoanalytically inspired interviews, in which the researchers used therapeutic techniques to circumvent their subjects' defenses in order to learn about their prejudices and authoritarian personality traits. In the interviews, the freedom of expression offered to the interview person was seen as the best way to obtain an adequate view of the whole person because it permitted inferences of the deeper layers of the subjects' personalities behind the antidemocratic ideology. An indirect interview technique, with a flexible interview schedule, consisted of "manifest questions," suggestions for the interviewer to pose in order to throw light on the "underlying questions," derived from the project's theoretical framework. These underlying questions had to be concealed from the subject so that undue defenses would not be established through the subjects' recognition of the real focus of the interview, namely to uncover the personally rooted causes of anti-Semitism.

The whole setup of the study was based on the realist idea that there is a "working mechanism" (a certain style of upbringing, which leads to a problematic personality structure) that—given the right context—leads to an outcome (in this case, the spread of fascism). Of course, the study—like all realist studies—is vulnerable to the objection that the latent or hidden mechanism that

> is said to generate the outcome could be illusory. Its validity in this case rests on the validity of psychoanalytic theory, which was widely accepted at the time of the study but much less so today. Furthermore, there are also potential ethical problems inherent in the ambition of "going beyond" people's self-understanding in order to uncover a truer reality behind their own experiences. I discussed this in Brinkmann and Kvale (2005), and, using an expression from Fog (2004), the application of psychoanalytic knowledge of defense mechanisms in the study by the Adorno group served as a "Trojan horse" to get behind the defense walls of the anti-Semites. When studying sensitive issues such as authoritarianism and anti-Semitism, it is difficult to obtain informed consent because this would likely render the study impossible. Thus, today, with our ethical standards, the Adorno project would be difficult, if not impossible, to carry out. In any case, it may here serve as an illustration of how a realist approach can look—with the stated goal of "going beyond" what is manifestly experienced to uncover the hidden structures and mechanisms that causally generate some phenomenon of interest. The method—interviewing—was in this case uncontroversial, but the theoretical interpretations that led to the production of knowledge were more debatable.

presented as a form of realism, but it is actually an anti-realism that reduces knowledge claims to what we may positively verify in experience. Causality consequently becomes constant conjunction in experience. In contrast to this, realist positions argue that science should go beyond immediate experience to study working mechanisms that generate the phenomena that we in fact experience. Philosophers today disagree about the existence of such mechanisms with regard to human psychological and social life. Some constructionists argue that there are no causally effective mechanisms in our social life (Harré (2002) thus tries to debunk what he calls the myth of social structure), whereas others, especially critical realists, argue that social science should be all about identifying such mechanisms.

I have tried to demonstrate in this chapter how positivism in qualitative research lives on in the practices informed by CAQDAS in particular (although it is certainly possible to work with these software programs without subscribing to positivism), whereas realism is influential in a number of other ways. Realism in particular employs the strategy of making the hidden obvious, and this can be seen as more than an intellectual commitment for, if Hadot is correct that philosophy should be considered as a way of life, it also represents a kind of existential stance to the world and other human beings. The notion that the world has "depth," and that other people live lives in their own right and are affected by forces that they might not be aware of, is important as a basic scientific credo and also as a moral commitment to uncover the hidden realities in order, perhaps, to help people with emancipation. This has been the agenda for the Frankfurt school of critical theory (Horkheimer, Adorno, Habermas, Honneth, etc.) and also for influential social scientists such as Pierre Bourdieu (1977). In that sense, the British traditions (as I call them here in a somewhat simplifying manner) live on in very significant ways in today's qualitative research communities.

GERMAN PHILOSOPHIES OF QUALITATIVE RESEARCH

PHENOMENOLOGY AND HERMENEUTICS

THE SCIENCES CAN BE considered as human practices that we engage in to produce knowledge in systematic ways. But the production of knowledge can take many different forms. Most people probably think of the sciences as explanatory: It is useful for humans to be able to explain how things work—plants, the continental plates, societies, the physiological system, and so on—in order for us to predict and control how events will unfold. But science also has a more purely descriptive task. This involves providing answers to questions about what exists. And when we are in the area of human lives and experiences, then the answers to the "what exists" question seem to be closely interwoven with how we experience that which exists. Thus, unlike anatomy, for example, which as a medical discipline rests on descriptions of the parts of the body (and not on how embodied creatures experience having a body), the human sciences rest on descriptions of the phenomena that we in fact experience. This, at least, is the starting point for phenomenology as a descriptive approach to qualitative human and social science.

Phenomenology sets itself the task of describing the essential structures of human experience. It is not about explaining the scientific or theoretical background to our experience (related to the working of the brain and central nervous system, for example), but it focuses directly on experience itself. This chapter is about phenomenology as a philosophy relevant for qualitative researchers, as it was developed in Germany from approximately 1900. It is also about phenomenology's close relative called hermeneutics, which historically predates the development of phenomenological philosophy proper but which is here treated after phenomenology because some of the more recent hermeneutic thinkers, such as Heidegger and Gadamer, engaged in critical discussions with the founder of phenomenology, Edmund Husserl. I begin this chapter by unfolding the original Husserlian phenomenology before moving on to what is known as existential phenomenology as it developed after Husserl in the hands of philosophers such as Heidegger, Sartre, and Merleau-Ponty. Especially Heidegger provides a bridge to hermeneutics—the art of interpretation—which adds an interpretive approach to the more purely descriptive ambitions of the phenomenologists. Hermeneutics already began with Schleiermacher's approach to biblical exegesis in approximately 1800, and this was approximately 100 years before Husserl developed his phenomenological philosophy. Both phenomenology and hermeneutics have had—and continue to have—immense influence on qualitative inquiry in many different forms.

Husserl's Phenomenology

The term *phenomenology* existed before Husserl. It literally means the study of the phenomena, with the understanding that the term *phenomena* refers to the ways in which objects and events appear to us. As Crowell (2009) notes, the term was used by Kant in the 18th century and by Hegel in the early 19th century, but because Husserl paid no attention to the previous uses of the term, there is no need to go back before Husserl if one wants to understand what he meant by phenomenology. Husserl's contributions were prolific and deep, ranging from his habilitation thesis from 1887 on the concept of number to his books from the 1930s and onward on the crisis in the European sciences (in which he introduced the celebrated idea of the lifeworld) and much else (Husserl, 1954) until his death in 1938. From Crowell's thorough chapter on Husserlian phenomenology, we can distill four main points that characterize his development of phenomenological philosophy:

First, "phenomenology is a *descriptive* enterprise, not one that proceeds by way of theory construction" (Crowell, 2009, p. 10). The goal of phenomenology is to provide descriptions of how we experience the world prior to the theories that we may formulate about it. Our emotions, perceptions, and thoughts are about something that appears to us in experience before we learn about its physical, chemical, biological, or psychological components. A depression, for example, is an experienced phenomenon prior to the theories we develop about it (neurochemical, psychoanalytical, or whatever). Phenomenology is not against theory building but, rather, sets itself the task of providing pre-theoretical descriptions. Of course, this has also provoked critical responses from scholars who deny that anything is "given" in our experience independently from theories and reflections (Sellars, 1956/1997). We saw the similar kind of discussion in relation to the phenomenalism of positivism in Chapter 3, in which the positivists were accused of denying the theory-ladenness of observation and of believing in the possibility of reducing our experience to elementary sense data. In general, however, the phenomenologists were not phenomenalists who believed that we can break experience down to simple sense data as components. On the contrary, they would insist that this is *not* how the world appears to us most of the time. Instead, it appears to us as always already infused with meaning, significance, values, and—perhaps—theoretical notions, but then it is *this* that should be described. In this sense, and unlike the positivists, the phenomenologists did not have a theory of what we experience (e.g., elementary sense data) but simply a respect for the many ways in which the world appears to us in experience.

Second, "phenomenology aims at *clarification*, not explanation" (Crowell, 2009, p. 10). Now, this is obviously related to the first point because a description of *how* we experience something is not the same as causally explaining *why* we experience something. Again, phenomenologists are not necessarily against causal explanation, but they insist that there is a descriptive step that should always be taken before it even makes sense to explain something causally (one must first know *what* to explain). I add, however, that much of Husserl's work consisted of critiques of *psychologism*— that is, the philosophical theory that logic can be explained with reference to how humans actually think and reason psychologically (in other words, that logic is founded on psychology).

Husserl reacted against this because it would mean reducing the *normativity* of logic to *causal* explanations of how the psychological system works. And more generally, there was in Husserl's phenomenology an awareness of the normativity of our experience. Intentionality was a key concept for him, which he took from Brentano. It is common to characterize mental life by saying that intentionality is the mark of the mental. It means that experience is always *about* something—our thoughts, feelings, perceptions, and actions are always directed at something. But, as Crowell states, "Intentionality is not simply the static presence of a 'presentation' in a mental experience (*Erlebnis*) but a normatively oriented *claim to validity*" (p. 13). In colloquial terms, this means that what we experience can only be "about" something (intentionality) because there are more and less correct and valid ways of experiencing it (normativity). For example, we may see a dangerous snake in the forest, but—on closer scrutiny—it may turn out to be an innocent branch, and our intentional orientation toward the object involves a normative underpinning of trying to "get it right." Simply stated, it means that we experience normativity, valences, and values in objects and events—something the Gestalt psychologists took up after Husserl (Köhler, 1938/1959).

Third, "phenomenology is an *eidetic* and not a factual inquiry" (Crowell, 2009, p. 10). This presents the most significant problem in relation to the development of phenomenology as a qualitative research practice. For, as Crowell adds, "Phenomenology studies some concrete act of perception only as an example for uncovering what belongs necessarily to perception as such" (p. 10). In other words, the goal is not to describe how Jack and Jill experience depression but, rather, to come to understand what are the essential properties of the experience of depression. In Husserl's version, phenomenology is thus wedded to a rather strong form of essentialism: There is something about depression that *makes* it depression, regardless of who has the experience—otherwise it would not make sense to call it depression. The goal is to obtain an understanding of the *eidos* of the experience—that is, its essential nature. A key procedure in this regard is what Husserl called free variation in one's imagination of the kind of phenomenon that one studies. That which one cannot remove from the phenomenon without turning the phenomenon into something else is its *eidos*.

Later, a concrete attempt at developing this idea into a qualitative research method is presented.

Fourth, "phenomenology is a *reflective* inquiry" (Crowell, 2009, p. 10). This is another way of saying what I highlighted previously: Phenomenologists do not deal directly with entities (planets, rivers, or bodies) but, rather, with our experience of entities. One should be careful, however, not to interpret this as a kind of mentalism or solipsism (solipsism is the philosophical idea that only the conscious contents of my experience exist). For the principle of intentionality should ensure that experience is always already pointing away from itself and toward worldly objects and events. Husserl thus subscribed to an anti-representationalist theory of the mind (p. 16), according to which the mind is *not* a container in which Cartesian ideas are experienced but, rather, something extending outward, so to speak, to the objects that are experienced. This became even clearer with Heidegger's phenomenology of practical engagement in the world and Merleau-Ponty's phenomenology of the living, acting body.

Existential Phenomenology

Husserl's philosophy was in many respects an attempt to break with a certain kind of naturalism and empiricism, and also with representationalism. Thus, he discarded the notion of the mind as a container into which contents from the outer world should be poured with the aid of the proper positivist methods. The mind is always intentional and normative. But the philosophers who later built on Husserl each tried to correct his philosophy in different ways. They did not think that he had broken sufficiently with the Cartesian philosophy that went before. I shall here introduce the leading post-Husserlian phenomenologists, whom I refer to as existential phenomenologists, because they were not concerned with Husserl's starting point in the transcendental ego (the subjective, experiential pole that is implicit in and prior to all experience) but, rather, in existence as such. They thus attempted to articulate existential conditions for experience—for example, in practical engagement (Heidegger) or the body (Merleau-Ponty)—representing what philosophers now call the "background" that is a precondition for experience.

Heidegger is a deeply problematic philosopher, especially for political reasons, because he was an awoved Nazi during Hitler's regime in Germany, and he never really went against this in later life. Despite this, however, many philosophers consider him one of the greatest philosophers of the 20th century, provided one can separate his political views from his philosophical ones. In his magnum opus from 1927, *Being and Time*, Heidegger set himself the task of describing what it means to understand or, more precisely, the entity that understands (Heidegger, 1927/1962). This was supposed to be preparatory work before developing a philosophy of being as such. Heidegger argued that philosophers in the West had forgotten the original philosophical question of being and instead merely asked about beings (i.e., What exists?). He lamented that we no longer inquire into the meaning of being (the ontological question) but only about the entities that actually exist (the ontic question). He thought it was necessary to more or less destroy the Western philosophical tradition (on this point, he found inspiration in Nietzsche) in order to be able to formulate the basic ontological question again. The destructive ambition (in an intellectual sense) in Heidegger's writings also explains why he became a source of inspiration for the later deconstructionist wave of thinkers in France (notably Derrida), who are discussed in Chapter 6.

Heidegger's name for the entity that understands is *Dasein*, and the being of *Dasein* is unlike the being of other entities in the universe because it alone can open for an understanding of being. Physical entities such as molecules, tables, and chairs are things that have categorical ontological characteristics, whereas human beings or *Dasein* are *histories* or *events* and have existentials as their ontological characteristics (Polkinghorne, 2004, pp. 73–74). The existentials that make up the basic structure of *Dasein* are *affectedness* (*Befindtlichkeit*) (we always find ourselves "thrown" into situations in which things already matter and affect us), *understanding* (*Verstehen*) (we can use the things we encounter in understanding the world), and *articulation* or *telling* (*Rede*) (we can to some extent articulate the meanings things have) (Dreyfus, 1991). In short, humans are creatures that are affected by what happens, can understand their worlds, and communicate with others. Of course, humans also have living physical bodies with categorical characteristics; we are *alive* as bodies and organisms, but in addition we have *lives* that unfold in time and evolve around what is deemed meaningful and valuable. So *Dasein* primarily exists as involved in a world of meanings, relations, and purposes and only derivatively in a world of objectified properties. In our everyday lives, we live absorbed in pre-established structures of significance, or what Heidegger called *equipment*, which serve as a background that enables specific things to show up as immediately meaningful and value-laden, given our participation in different social practices. This is close to Husserl's concept of lifeworld as the world we experience prior to reflection or theorization. We do not experience meanings and values as something we subjectively project unto the world, for the qualitative world in which we live meets us as always already imbued with meaning and value. Only when our everyday, unreflective being-in-the-world breaks down, when our practices of coping with the equipment somehow become disturbed, do entities appear with "objective" characteristics distinguishable from human subjects. Thus, for Heidegger, it was central that humans primarily meet the world in a mode he called "ready-to-hand," in which we are practically engaged. In this mode, things show up as "worldly entities" (Wrathall, 2009, p. 33)—that is, appearing as "affordances" that invite us to grasp, move, and use them in designated ways. What Heidegger called "presence-at-hand" is a derivative way of relating to the world, in which things become detached objects with discrete properties that may be studied scientifically or theoretically. Worldly things, Wrathall says, "are structured by the ways that they relate to and condition other things, bodies, and activities" (p. 36). A hammer is not just a hammer in isolation but appears as such within a network of things (nails, wood, etc.) and activities (carpentry, building, etc.). The world itself is an existential phenomenon (p. 37) that should be described phenomenologically, including the anxiety that humans may feel when confronted with the fact that the way the world is given in experience is contingent and not metaphysically guaranteed. Like Kierkegaard, Heidegger would thus say that existential anxiety shows us nothingness and therefore that things could be different.

As Dreyfus and Taylor (2015) make clear in a book in which they advance an existential phenomenological outlook, the basic phenomenological approach (at least in Heidegger and Merleau-Ponty) is *realist*. It is not realist in a representationalist sense, where we can *copy* the world in our representations, but rather in a pragmatic sense, where we can *cope* more and more effectively with the world (p. 142). Through practical coping, we also get a better "conceptual grasp of the objects that surround us," which means that "some accounts of nature provide a better explanation of how the universe works than do others" (p. 142). They thus find in phenomenology a critique of the epistemological tradition that posits a split between subject and object, and instead they argue that our first understanding of reality "is not a picture I am forming of it, but the sense given to a continuing transaction with it" (p. 70). The philosophical riddles (How can I know anything?) arise when we forget that we have a preconceptual and prereflective familiarity with the world. Merleau-Ponty's (1945/2002) philosophy demonstrated how this familiarity springs from our living, moving, and sensing bodies. Through bodily practices and our "motor intentionality," we inhabit and know the world in a primitive and important way—this is where lifeworld phenomena are grounded:

> All my knowledge of the world, even my scientific knowledge, is gained from my own particular point of view, or from some experience of the world without which the symbols of science would be meaningless. The whole universe of science is built upon the world as directly experienced, and if we want to subject science itself to rigorous scrutiny and arrive at a precise assessment of its meaning and scope, we must begin by re-awakening the basic experiences of the world of which science is the second order expression. (p. ix)

This is Merleau-Ponty's way of saying that the ready-to-hand is more basic than the presence-at-hand perspective we may gain through scientific activity. Using a metaphor, we can say that when we are concerned with how humans live and experience their lives, the sciences may give us maps, but the practical world of everyday life is the territory or the geography of our lives. Maps make sense only on the background of the territory, where human beings act and live, and should not be confused with it. Phenomenologists are not against scientific abstractions or "maps," but they insist on the primacy of concrete descriptions of experience—of that which is prior to maps and analytic abstractions. Dreyfus and Taylor (2015) summarize and present some fundamental insights to be gained from Heidegger and Merleau-Ponty: First, as human scientists, we need to focus on a kind of understanding that is preconceptual;

second, that understanding is related to the human being as an engaged agent; and, third, the human being as an engaged agent is in "bodily commerce with our world" (p. 69). But there is also a fourth point, emphasizing the fact that "our humanity also consists in our ability to decenter ourselves from this original engaged mode—to learn to see things in a disengaged fashion, in universal terms" (p. 69). This is how phenomenologists find room for scientific practice as a universalizing practice of abstraction—not as something that replaces our embodied, engaged understanding of the world but, rather, as a mode of relating to the world that we may jump into from time to time because it is useful for certain purposes. But these purposes spring from the lifeworld, which is therefore the ultimate locus of meaning and significance. Thus, whereas the positivists had seen scientific practice as epistemologically basic—as the ground of all genuine human knowledge—the existential phenomenologists see science as a special kind of universalizing practice that grows out of more primordial lifeworld activities.

There are other influential phenomenologists, not least Jean-Paul Sartre, who developed an existentialism proper, which depicted the human being as fundamentally free (Sartre, 1943/1966). According to Sartre, our consciousness is a being-for-itself, which means that we are always already ahead of ourselves in a way, projecting ourselves into the future by way of free choices. This is what existence means, which Sartre contrasts with the kind of being-in-itself that characterizes objects in the world. They merely are, so to say, and do not exist (the etymological root of "existence" is to "stand out" rather than merely stand). A famous existentialist slogan from Sartre is that existence precedes essence. This means that there is no predetermined essence in the human being. Even if we have both biological and cultural heritages, these do not determine us, Sartre claimed. We create ourselves through free and radical choice—"radical" because we do not simply choose between A and B, but we are also free to choose that this very choice is a significant one to make. This is too radical for most other phenomenologists, and Taylor (1985b) has articulated an influential critique of Sartre on this point, seeking to show that this would undermine the very idea of choice that seems to presuppose something unchosen. Furthermore, it seems plain wrong, phenomenologically, to say that we are basically such freely choosing agents, when there is so much unchosen "facticity" in our lives (as Heidegger would say). The late works of Sartre tried to incorporate these matters into his existentialism, not least through dialogue with Marxism in his *Critique of Dialectical Reason* first published in 1960.

I emphasize one last figure, who is not conventionally listed as a phenomenologist. I am thinking of Ludwig Wittgenstein, arguably the greatest philosopher of the 20th century. However, I believe we have reason to follow Gier (1981) and read Wittgenstein's work phenomenologically. He was not a phenomenologist of experience but, rather, of language. Wittgenstein shared with Husserl and the other phenomenologists an ambition of describing. In *Philosophical Investigations* (Wittgenstein, 1953) and other key works, he showed that philosophical problems can be dissolved through careful descriptions of how we actually use the concepts in our language. Like Merleau-Ponty, Wittgenstein was interested in language as an embodied, practical activity, and like Heidegger, he was interested in bringing the non-thematized background to the fore, which is what enables our sayings and doings to be meaningful. If I say "Please give me the key," then this sentence is meaningful not just because of a structural relationship between the different parts of the sentence and the world but also—and more importantly—because of a nonreflective background in which there are locks, keys, and certain actions involving opening and closing doors, and so on. Wittgenstein's "linguistic phenomenology" thus succeeded in showing how the meanings of our words are found in the practical uses to which they are put. Meaning is not connected to "mental representation" in our heads but, rather, to social practice. Although Wittgenstein's philosophy has mainly made a stamp on analytical philosophy in the United States and the United Kingdom, I believe we have good reasons to approach him as a philosopher, who was interested in the same kinds of questions as the phenomenologists. And, in many ways, he gave the same kinds of answers, namely that we must look to the lifeworld—the world as actually lived, experienced, and talked about—if we want to understand the human being.

Just as the other phenomenologists, Wittgenstein certainly employed the philosophical strategy of "making the obvious obvious" to invoke the phrase from Noblit and Hare (1988). Malcolm's wonderful book about Wittgenstein bears the significant title *Nothing Is Hidden* (Malcolm, 1988). That nothing is hidden is a philosophical insight, according to Wittgenstein. Of course, sometimes our keys are gone because they are hidden underneath the pillows, but when we talk about philosophical problems, we really talk about something that should be understood by looking and describing the world and our language use in particular—and *not* by seeking explanatory mechanisms beyond what actually appears to us. This, I believe, is the basic phenomenological outlook: That nothing of philosophical interest is hidden, but that significant phenomena can be so obvious to us—so implicit in our sayings, doings, and experiences—that we fail to notice them. The task of phenomenology is to look again, to re-search, and describe—not primarily the maps or other abstractions that we have created in our scientific endeavors, however important these may be, but the actual territory of the lifeworld.

Hermeneutics

Hermeneutics is in many ways related to phenomenology and is the other main tradition that originates in Germany but which—like phenomenology—has become influential on a global scale. Simply stated, hermeneutics is the art of interpretation. On some accounts, this art can be formalized into a specific method (that can be used, for example, by qualitative researchers), but most hermeneutic scholars maintain that interpretation is not so much a specific method as it is a manner of being in the world as such. Originally, with Friedrich Schleiermacher, hermeneutics was developed as a methodology for interpreting texts, notably biblical texts (what follows builds on Brinkmann, 2012). Unsurprisingly, it is quite a challenge to develop a methodology for uncovering the meaning of texts that were written hundreds of years ago in a different context. Of course, theological, philosophical, and legal scholars had been concerned with interpretation before Schleiermacher, but he is considered to be the father of hermeneutics because of his focus on the act of interpretation itself. Schleiermacher's focus was on *text* interpretation specifically, but with Wilhelm Dilthey in the late 19th century, hermeneutics was extended to human life itself, conceived as an ongoing process of interpretation. All human acts, objects, and expressions were seen as demanding interpretation to be understood.

In 1894, Dilthey (1894/1977) developed a hermeneutic "descriptive psychology," portraying life as an ongoing process of interpretation lived on the background of historical meanings. It was therefore Dilthey's idea that the methods of the *Geisteswissenschaften* (the moral sciences) were different from those of the natural sciences because the historical and social world in which we live is grasped by *understanding* and *interpretation* rather than *explanation* and *measurement*. With Heidegger's (1927/1962) *Being and Time* from the early 20th century, we see another shift from an epistemological hermeneutics of life to ontological hermeneutics proper. The question of methodological hermeneutics (e.g., Schleiermacher) had been: How can we correctly understand the meaning of texts? Epistemological hermeneutics (e.g., Dilthey) had asked: How can we understand our lives and other people? But ontological hermeneutics—or "fundamental ontology" as Heidegger also called it (p. 34)—prioritizes the question: "What is the mode of being of the entity who understands?" (Richardson, Fowers, & Guignon, 1999, p. 207). As we have seen, *Being and Time* aimed to answer this question and can thus be said to be an interpretation of interpreting, or a philosophical anthropology (a philosophy of the *anthropos*, the human being).

For Heidegger and later hermeneuticists such as Gadamer and Taylor, understanding is not something we occasionally do—for example, by following certain procedures or rules. Rather, understanding is the very condition of being human (Schwandt, 2000, p. 194). We always see things *as* something, human behavior *as* meaningful acts, letters in a book *as* conveying some meaningful narrative. In this way, the hermeneutic addition to phenomenology can be said to concern how our directedness or intentionality is always an *interpreted* intentionality. Our directedness to objects and events is necessarily an interpretive act. This is not to say that we can freely choose how we direct ourselves in acts of interpretation—we are certainly constrained by history, culture, and what Gadamer called our prejudices—but it is to say that there is no description that is entirely free from interpretation—implicitly or explicitly. In a sense, then, understanding something is an act, and hermeneutic writers argue that all such understanding is to be thought of as interpretation. This, however, should not be understood as implying that we normally make some sort of conscious mental act in interpreting the world. "Interpretation" here is not like the mental act of interpreting a poem, for example. It is not an explicit, reflective process but, rather, something based on skilled, everyday modes of comportment (Polkinghorne, 2000). Our reflective ways of interpreting the world are grounded in tacit, embodied ways of being in the world.

Interpretation thus depends on certain *prejudices*, as Gadamer (1960/2000) famously argued, without which no understanding would be possible. Knowledge of what others are doing and knowledge of what my own activities mean "always depend upon some background or context of other meanings, beliefs, values, practices, and so forth" (Schwandt, 2000, p. 201). There are no fundamental "givens," for all understanding depends on a larger horizon or background of non-thematized meanings. This horizon is what gives meaning to everyday life activities, and it is what we must engage with as we do qualitative inquiry; both as something that sometimes breaks down and necessitates a process of inquiry and as something that we can try to make explicit in an attempt to attain a level of objectivity. The latter exercise is often referred to by qualitative methodologists as making one's pre-understanding explicit.

According to the hermeneutic anthropology—its image of human beings—we are self-interpreting animals (Taylor, 1985a) who make sense of the world through storylines and interpretative repertoires that generate meanings. The writer of fiction, Douglas Coupland (2009), states it well through one of his characters in the novel *Generation A*: "How can we be alive and not wonder about the stories we use to knit together this place we call the world? Without stories, our universe is merely rocks and clouds and lava and blackness" (p. 1). Our human lives—actions, thoughts, and emotions—are nothing but physiology if considered as isolated elements outside of narrative and interpretative contexts. A life, as the great French hermeneutic philosopher Paul Ricoeur (1991) has said, "is no more than a biological phenomenon as long as it has not been interpreted" (p. 28). That we are self-interpreting creatures that use stories to constitute our identities and as means of self-understanding (Brinkmann, 2008) does not imply that any individual person can make himself Napoleon by trying to interpret himself that way—that is, by telling the story that he is Napoleon—but it means that human communities of interpreters and their historical and narrative traditions constitute the meanings of our self-interpretations. Gadamer (1960/2000) stated,

> In fact history does not belong to us; we belong to it. Long before we understand ourselves through the process of self-examination, we understand ourselves in a self-evident way in the family, society, and state in which we live. The focus of subjectivity is a distorting mirror. The self-awareness of the individual life is only a flickering in the closed circuits of historical life. *That is why the prejudices of the individual, far more than his judgments, constitute the historical reality of his being.* (pp. 276–277)

Gadamer (1960/2000) argues that our nature as "self-interpreting animals" makes the condition of human and social science quite different from the one we find in the natural sciences, "where research penetrates more and more deeply into nature" (p. 284). In the human and social sciences, there can be no "object in itself" to be known (p. 285), for interpretation is an ongoing and open-ended process that continuously reconstitutes its object. The interpretations of social life offered by researchers in the human and social sciences are an important addition to the repertoire of human self-interpretation, and influential fields of description offered by human science, such as psychoanalysis, can even affect the way whole cultures interpret themselves. As Richardson and colleagues (1999) spell out the consequences of Gadamer's view, this means that "social theories do not simply mirror a reality independent of them; they define and form that reality and therefore can transform it by leading agents to articulate their practices in different ways" (p. 227). As Taylor has argued throughout the years, human and social theory is therefore best thought of as a kind of practice, whose validity should not be judged with reference to how well it represents the world as such but, rather, with reference to its capacity to enrich and improve the phenomena and practices under consideration. If the prejudices of the individual are what constitutes the human being, as Gadamer stated in the previous quote, it also implies that any understanding of social processes runs in two directions at the same time: toward the social, with an eye to the historical and cultural processes that form our habits and prejudices, but also toward the one who lives in the social and cultural world—the person of the researcher—who must take her own biography (and prejudices) into account and do her analytic work in the intersection between the personal and the social, as C. Wright Mills (1959/2000) advocated in his call for the sociological imagination. This is where we can find a hermeneutic springboard for qualitative inquiry.

Before moving on to consider qualitative research applications of phenomenological and hermeneutic philosophies, we can sum up concerning their perspectives on scientific practice. Husserl wanted phenomenology to be a strict descriptive science of the structures of experience, but already with his followers, Heidegger and Merleau-Ponty, this became more complex. Especially Heidegger paved the way for modern hermeneutics, which was continued by his student Gadamer. For Gadamer, the human and social sciences are closer to "moral knowledge" than "theoretical knowledge." As Gadamer (1960/2000) stated,

> The human sciences stand closer to moral knowledge than to . . . "theoretical" knowledge. They are "moral sciences." Their object is man and what he knows of himself. But he knows himself as an acting being, and this kind of knowledge of himself does not seek to establish what is. An active being, rather is concerned with what is not always the same but can also be different. In it he can discover the point at which he has to act. The purpose of his knowledge is to govern his *action*. (p. 314)

Gadamer thus views knowledge in the social sciences as *practical* knowledge, and he emphasizes the dimension of *application* in all human understanding. Hermeneutics is therefore *practical philosophy*, and Gadamer refers to Aristotle's conception of practical wisdom—*phronesis*—as the model for hermeneutic understanding (cf. the discussion in Chapter 2). Understanding is always bound up with application, and in *Truth and Method*, Gadamer develops this view from legal hermeneutics, where understanding the law necessarily involves applying it to concrete cases. One does not really understand the law if one is unable to apply it. New applications again develop the practice of law creatively, just as future instances of legal judgment must refer back to earlier instances—that is, to the *tradition* of law. And so it is with human science in general, Gadamer argues. In understanding what humans do, we apply certain historically evolved descriptions to their activity, and as acting and self-interpreting beings, humans can then employ social researchers' understandings as self-understandings (Brinkmann, 2011). This has been called a triple hermeneutics: Qualitative researchers interpret a human reality that is always already constituted by interpretations, and researchers' interpretations may feed back into this reality.

Phenomenology and Hermeneutics as Qualitative Research Practices

As discussed previously, the short story of phenomenology begins with Husserl's interest in experience and the lifeworld and is continued with the existential philosophies of Heidegger, Sartre, and Merleau-Ponty. The goal in Husserlian phenomenology was to arrive at an investigation of essences—that is, to describe the essential structures of human experience from a first-person perspective. How can this be developed into a qualitative research practice?

The implications of phenomenological philosophy for qualitative research were first developed in a series of studies at Duquesne University (Brinkmann & Kvale, 2015). Starting with van Kaam's (1959) early study of "the experience of really being understood," the method was further applied, systematized, and reflected upon by the phenomenological psychologist Amedeo Giorgi and co-workers (Giorgi, 1975). According to Giorgi, "Phenomenology is the study of the structure, and the variations of structure, of the consciousness to which any thing, event, or person appears" (p. 83). The goal is to arrive at an investigation of essences by shifting from describing separate phenomena to searching for their common essence. As discussed previously, Husserl termed one method of investigating essences a "free variation in fantasy," which means varying a given phenomenon freely in its possible forms, and whatever remains constant through the different variations is the essence of the phenomenon. Another important methodological concept is the phenomenological *reduction*. This signals a call for a suspension of the researcher's judgment as to the existence or nonexistence of the content of an experience. The reduction is often referred to as a "bracketing," an attempt to place the common sense and scientific foreknowledge about the phenomena within parentheses in order to arrive at an unprejudiced description of the essence of the phenomena.

To give an example, a particularly well-crafted phenomenological study is "Throwing Like a Girl: A Phenomenology of Feminine Body Comportment, Motility and Spatiality," by political scholar and feminist Iris Marion Young (1980; also discussed in Brinkmann, 2012). In her phenomenological analysis, Young addressed the everyday activity of throwing objects. She began by quoting the phenomenologist Erwin Straus, who, in an earlier essay from 1966, had observed what he called a "remarkable difference in the manner of throwing of the two sexes" (p. 137). Girls at the age of 5 years, Straus had noted, does not make any use of lateral space when throwing an object such as a ball. A boy of the same age, on the other hand, uses his entire body when throwing, which results in a much faster acceleration of the ball. Straus ascribed this observed difference to an innate biological difference, a specific "feminine attitude" that girls have in relation to the world and to space that is not as such anatomical but, rather, biological at some deeper level. This difference between boys and girls results in the derogatory expression "he throws like a girl," which is used, for example, as an insulting comment in American baseball. Young finds some of Straus' descriptions compelling, but she disagrees with his explanation, which invokes "a mysterious feminine essence" (p. 138) in order to account for the difference between boys and girls. So Young significantly expands on Straus' description through an extremely close portrayal of the motions of gendered bodies. Young states,

> Girls do not bring their whole bodies into the motion as much as the boys. They do not reach back, twist, move backward, step, and lean forward. Rather, the girls tend to remain relatively immobile except for their arms, and even the arm is not extended as far as it could be. (p. 142)

Young goes on to describe further differences in bodily comportment between boys and girls, and also men and women, concerning how they sit, stand, and walk:

> Women generally are not as open with their bodies as men in their gait and stride. Typically, the masculine stride is longer proportional to a man's body than is the feminine stride to a woman's. The man typically swings his arms in a more open

and loose fashion than does a woman and typically has more up and down rhythm in his step. (p. 142)

There are many other rich descriptions in Young's essay—for example, concerning how women tend to lift heavy objects:

Women more often than men fail to plant themselves firmly and make their thighs bear the greatest proportion of the weight. Instead, we tend to concentrate our effort on those parts of the body most immediately connected to the task— the arms and shoulders—rarely bringing the power of the legs to the task at all. (p. 143)

Young's (1980) prose is precise, descriptive, and dry, but she manages to take the mundane movements of girls and women and present them as a very relevant and interesting topic for analysis. As a reader, one is convinced by her accurate descriptions; there is an immediate feeling of recognition. In a sense, we know already what Young is saying, but we do not know that we know it, and in that way, her article is a good example of successfully applying the phenomenological strategy of making the obvious obvious for us, something that is extremely difficult to do in such an insightful way. Note that Young does not stay with pure description, for she also argues that women are holding themselves back, not just concerning their bodily movements but also more generally in society. Unlike boys, who learn to be assertive, aggressive, and proactive, and grow up to act in this way as adult men, girls are not taught to use their full bodily capacities "in free and open engagement with the world" (p. 152). Consequently, they often become insecure, restricted, and self-enclosed as adult women. This, for Young the feminist scholar, is not to be explained with reference to anatomy, physiology, or by invoking a mysterious feminine essence but is rather a result of the particular situation of women in a patriarchal and thus oppressive society. This means that questions about gender inequality turn out not just to be a matter of discourses and the symbolic (e.g., concerning unequal rights and treatments). Gender inequality may be rooted in the very bodily habits that persons acquire from a very early age, which may be quite difficult to change. A first step toward improvement, however, can be taken when these inequalities are made obvious for us. Furthermore, Young does not stay with the merely descriptive. She also invokes theoretical concepts to present an explanation of the phenomenon that she has observed. And we can even say that her initial descriptive stance is itself theoretically informed, namely by the phenomenology of the body as developed by Merleau-Ponty (1945/2002). Young's (1980) analysis takes observations of men and women, boys and girls, as the starting point, but phenomenologists also often use interview materials in their analyses. In Brinkmann and Kvale (2015), there is a discussion of a concrete methodological approach called "meaning condensation," which is inspired by Giorgi (1975). Giorgi distinguishes between "natural units," which are segments of meaningful talk articulated in a qualitative research interview, and "central themes," which are condensations of the meanings articulated. By condensing the description into central themes, and by employing the reduction and bracketing of the themes, Giorgi argues that it becomes possible to arrive at phenomenological essences (e.g., the core structure of the experience). This is probably as close as we come to a specific phenomenological method for qualitative researchers (for further examples and discussion, see Langdridge, 2007). Box 4.1 presents an example of this method being applied in a concrete research project.

Unlike such phenomenological procedures, it is difficult to find proper hermeneutic guidelines for qualitative analysis. The reason probably is that since Gadamer there has been suspicion among interpretative scholars that it is possible to "codify" understanding and interpretation into discrete steps or procedures. Understanding is not a "method," and it is often rightly said that Gadamer's *Truth and Method* (1960/2000) should have been called *Truth or Method* for the title to have been more in line with the contents. Instead of providing a hermeneutical method, Brinkmann and Kvale (2015) referred to certain "canons of interpretations," which are more or less general hermeneutic frames—that is, aspects to think about when conducting a hermeneutic analysis. The following text is adapted from Radnitzky (1970) and also discussed in Brinkmann and Kvale (2015).

The first hermeneutical canon involves a continuous back-and-forth process between parts and whole, which follows from the so-called hermeneutical circle (where parts and whole reciprocally define each other). Starting with an often vague and intuitive understanding of the text or phenomenon as a whole, its different parts are interpreted, and out of these interpretations the parts are

Box 4.1 **Phenomenology and How to Make the Obvious Obvious**

The following is an example of a phenomenological study of anxiety as experienced in social situations, based on three individual descriptions (Beck, 2013). In line with the descriptive phenomenological method in psychology developed by Giorgi, the goal was to provide an analysis discovering the general structure of social anxiety. Beck describes the approach under the headings Method, Subjects, Procedure, Analysis, and Results:

> *Method*: "Since phenomenology is conceived by Husserl as a descriptive science based upon intuitions of concrete givens, it cannot proceed in the same way as formal, exact, eidetic sciences. . . . Nor does the phenomenological procedure involve induction, because that would involve generalization after encountering a certain number of concrete manifestations. . . . For phenomenology, the essential characteristic has to be intuited (seen) and described. This "seeing" is aided by the use of free imaginative variation" (Giorgi as quoted in Beck, 2013, p. 187).
>
> *Subjects*: "Phenomenological psychological research, being the psychological study of phenomena as experienced, requires rich concrete descriptions of personal experiences for analysis. Because of the depth expected within those descriptions, no particular variables need be controlled for nor diverse populations represented. Rather than searching for correlations between certain seemingly related variables, a phenomenological approach focuses on how certain experiences generally occur to consciousness, so that relationships between the resulting essential constituents of the phenomenon in question can be better comprehended" (Beck, 2013, p. 188). Furthermore, "Subjects were chosen for this study on the simple basis that they had reported to have experienced anxiety while participating in a socially orientated situation on at least one occasion. It was requested that this experience be remembered in enough detail for a sufficiently rich

description of the experience to be given. It was also requested that the participants be available for clarification should any part of their description be determined too vague. On the basis of these simple requirements, three participants were chosen with no discretion given to any other variables" (p. 188).

Procedure: "The participants were asked to report descriptions of personal accounts of anxiety that they experienced during a socially orientated situation. It was made clear that a single occurrence of distress experienced within a socially orientated situation, which the participant would classify as anxiety, would sufficiently fulfill the needs of this particular study" (Beck, 2013, p. 188).

Analysis: To begin the analysis, the phenomenological psychological attitude was assumed in order to more fully comprehend the meanings presented by the descriptions in their "own appropriate mode of self-givenness, thus [meeting] the demand for scientific objectivity concerning the subjective: the method of phenomenological reduction" (Scanlon as quoted in Beck, 2013, p. 189). Furthermore, "The final structure obtained from the three descriptions should ideally represent the general manner in which the experience of anxiety felt during social situations is lived through" (p. 189).

Results: The author reviews the meaning units obtained through the phenomenological reduction and arrives at a single structure, which depicts the experience of anxiety in social situations. I quote Beck's (2013) statement of the result here at length to illustrate the phenomenological commitment to obtaining an essential structure of an experience:

For P (the ideal participant), during an experience of anxiety in a social situation, a desire to fulfill a certain set of self-imposed arbitrary prescriptions transpires simultaneously alongside a feeling of self-lack to be fully present to the task at hand, due to self-perceived deficiencies that P feels compelled to keep hidden from potentially evaluating others. This awareness of P's own deficiencies is perceived by

> P as being evaluated by others, but is shown merely to be P's immanent awareness of her own ambivalence. Certain evaluations associated with the self-perceived deficiencies become attributed to any person included within the situation at hand, which leads P to perceive all persons included in the situation as one collective, threatening (feared) entity capable of critical evaluation rather than experiencing the uniqueness of each individual personality presented to him/her. The aforementioned ambivalence results in a duality of consciousness leading to dissonance for P, thus limiting P's ability to think rationally throughout the emotional instability. P proceeds to reflect on former perceptions related to emotions and thoughts experienced across various situations which have invoked similar perceptions in the past. P remains distracted by the internal conflict occurring between the two seemingly irreconcilable desires/intentions to the extent that P is unable to expose his/her true personality and naturally engage in the situation as hoped. No longer able to confidently trust his/her own perceptions, P's perceived role in relation to what becomes represented by the task at hand remains uncertain, potentially leading to further provoking experiences in future situations deemed similar in nature to the one in question. (pp. 190–191)
>
> Beck refers here to the "ideal participant" to underscore that the goal of this kind of phenomenological analysis is not to capture the experiences of these concrete individuals involved but instead to obtain a description that ideally applies to any instance of the experienced phenomenon, here anxiety in social situations. This demonstrates phenomenology's allegiance to a certain kind of scientificity and also how the researcher is interested in making what is implicit in an experience of a given phenomenon explicit—thereby making the obvious obvious.

again related to the totality, and so on indefinitely. We never arrive at the one true interpretation of something because all interpretative activity takes place in a historical situation, which is constantly changing. In the hermeneutical tradition, the circularity of parts and whole is not viewed as a "vicious circle" but, rather, as a "circulus fructuosis" or spiral, which implies a possibility of a continuously deepened understanding of meaning. The problem is not to get away from the circularity in the explication of meanings but, rather, to get into the circle in the right way.

A second canon is that an interpretation of meaning ends when one has reached a "good Gestalt," an inner unity of the text, which is free of logical contradictions.

A third canon is the testing of part-interpretations against the global meaning of the text and possibly also against other texts or statements by the same author or speaker.

A fourth canon is the autonomy of the text or utterance: The text should be understood on the basis of its own frame of reference, by explicating what the text itself states about a theme. Most hermeneuticists are critical of relativizing the meaning of something to the psychological state of the author or speaker. The goal is not to answer the question, "What did the person really mean?" but, rather, to come to an understanding of what meaning the text articulates in itself.

A fifth canon of the hermeneutical explication of a text concerns knowledge about the theme of the text. Knowledge is not viewed as hindering the process of interpretation but, rather, as an aid in understanding. Qualitative researchers should keep an open mind about their phenomena when trying to understand them, but, as some people say, an open mind is not the same as an empty head. In fact, it is quite often the case that a head full of experience and theoretical concepts is better equipped to interpret something because a person with such a head has many tools available for understanding, for example, what Herbert Blumer once called "sensitizing concepts" that make the interpreter receptive to specific meanings. This is quite close to Gadamer's argument about the necessity of prejudices for qualified judgments to be made (or foreknowledge as a precondition of knowledge).

Related to this is the sixth principle, which states that an interpretation of a text is never without presuppositions. The interpreter cannot "jump outside" the tradition of understanding he or she lives in. The interpreter of a text may, however, attempt to make his or her presuppositions explicit and attempt to become conscious of how certain formulations of a question to a text already determine which forms of answers are possible. This is definitely a virtue in most qualitative research projects.

A seventh and final canon states that every interpretation involves creativity. An interpretation goes beyond the immediately given and enriches one's understanding by bringing forth new differentiations and interrelations in the text, extending its meaning. As discussed in Chapter 7, post-structuralists put much emphasis on this and the deconstructive idea that meaning is forever "deferred," as Derrida would say, which means constantly postponed, because something has meaning only in a given context, and if the context is constantly changing, so is meaning.

Perhaps the most methodologically explicit approach to interpretation in qualitative inquiry is interpretative phenomenological analysis, which unites phenomenological description and hermeneutical interpretation (Smith, Flowers, & Larkin, 2009). As an approach, it has been applied most often in health psychology and "generally aligns its approaches to qualitative inquiry with the distinctive subdiscipline of cognitive psychology" (Thorne, 2014, p. 105). In this sense, it is quite in line with Husserl's original ambition of creating a strict descriptive science of consciousness and less in line with a hermeneutic notion of human science.

Summary

In this chapter, I presented the phenomenological and hermeneutic philosophies that have been immensely relevant for qualitative research. Phenomenology began with Husserl and was continued by Heidegger and Merleau-Ponty and developed into tools for qualitative inquiry by scholars such as Giorgi. Hermeneutics dates back to Scheiermacher and Dilthey and was in a sense merged with phenomenology by Heidegger and brought up to date by Gadamer in particular. Many qualitative methodologies employ strategies from phenomenology and hermeneutics, which I believe can be condensed to the essential idea of making the obvious obvious. The difference between phenomenology and hermeneutics in their purer forms concerns the extent to which they view interpretation (rather than description) as a necessary component in making that which is implicit in an "obvious" way explicit. In Chapter 5, we move from Europe to the United States and consider the tradition that some commentators conceive of as an American version of phenomenology, namely pragmatism (for a reading of William James as a phenomenologist, see Herzog, 1995).

Returning to a foundational principle of this book as articulated in Chapter 1 through Hadot: Philosophy is not just a set of intellectual ideas but also a way of life. How can phenomenology affect one's way of life as a qualitative researcher? The answer is that a phenomenologist is committed to a shared human world on which we each have our different perspectives. Living and doing research as a phenomenologist means learning to suspend one's beliefs about how others experience something and patiently observing and listening to them. Whether or not one subscribes to the phenomenologists' notion of essences to be uncovered by description and free variation of the phenomena in one's imagination, we should learn from the "reduction" and "bracketing" of the phenomenologists. As the hermeneutic philosophers stressed, we cannot not know—that is, get rid of our prejudices, because these are also a precondition of something being significant and meaningful in the first place—but we can strive to set aside what we think we know in order for the phenomena to stand out as clearly as possible and as little clouded by our subjective viewpoints as possible. This, I find, is not just a valid epistemic principle but also a worthy moral ideal.

5

AMERICAN PHILOSOPHIES OF QUALITATIVE RESEARCH

THE PRAGMATISMS

THE BRITISH TRADITIONS of positivism and realism are about objectivity and strict methodological procedures in the former case and an ambition of moving beyond appearances in the latter. The German traditions of phenomenology and hermeneutics are about describing and interpreting the human lifeworld. But what about changing the world? This, of course, can be the outcome of all the traditions discussed so far, but mainly as a later phase that comes after the research process. In this chapter, I consider the pragmatist philosophies of qualitative research, which are about knowing the world *through* changing it. Thus, for pragmatists, knowledge *is* action and not something that follows after the process of knowing is over. Some qualitative paradigms such as action research are change-oriented in a quite concrete sense, but pragmatism is a more philosophical approach to the human being and our world that depicts the true as the useful.

It is often said that pragmatism is *the* uniquely (North) American contribution to philosophy. Some Continental European thinkers have previously frowned on the action-oriented ideas of pragmatism and believe that the emphasis on *doing* in pragmatism is too reminiscent of a cowboy-like mentality (shoot first, ask questions later). Relatively recently, however, pragmatism has been recognized globally as a significant school of philosophy and also by thinkers in Germany such as the sociologist Hans Joas (1996), in France including the schools of the new French pragmatists (Boltanski & Chiapello, 2005; Latour, 2005), and in many other places. The ideas of the founders of pragmatism are now being compared to phenomenology and hermeneutics. William James had close ties with phenomenological philosophy (Herzog, 1995), and John Dewey has also been read as a kind of phenomenologist (Kestenbaum, 1977). Heidegger's existential phenomenology specifically has been coupled quite closely with pragmatism (Okrent, 1988). Dewey reportedly said that Heidegger's (1927/1962) *Being and Time* was just a version of his own magnum opus on man's being in the world, *Experience and Nature* (Dewey, 1925), except in "high-falutin' German" (May, 2009, p. 22). In this chapter, I introduce the philosophy of pragmatism, with a special focus on John Dewey, before recounting how qualitative sociology in particular, but also other fields of qualitative inquiry, has been influenced by pragmatist ideas.

What Is Pragmatism?

In his book on the history of pragmatism, Louis Menand (2002) has aptly characterized pragmatism as a single idea that was shared among its founders: Charles Sanders Peirce, William James, John Dewey, and also the (philosophically less known) supreme judge Oliver Wendell Holmes. This is an idea about ideas: "Ideas are not 'out there' waiting to be discovered, but are tools—like forks and knives and microchips—that people devise to cope with the world in which they find themselves" (p. xi). In many ways, this pragmatist idea about ideas was, and is, a revolutionary proposal that turns Western thought on its head. Ideas are not representations or copies of how the world is but, rather, are tools with which we transform, engage with, and cope with the world as we go about living our lives (in the following discussion, I have reworked some passages from Brinkmann, 2013a).

Pragmatism is thus more than yet another set of philosophical ideas; it represents a more fundamental shift in perception. The philosopher Stephen Toulmin (2001) supplements and characterizes Dewey's pragmatism in particular as "a change of view, which puts theorizing on a par with all other practical activities" (p. 172).

As Dewey himself liked to state, we need to understand the difference between theory and practice as a difference between two kinds of practice (Dewey, 1922/1930, p. 69). The act of theorizing, of formulating ideas concerning the world, is just as much a kind of practice as is building a house, caring for children, or writing computer programs. All of these are activities by which people seek to gain a foothold in a changeable world. As human beings, we exist in this world fundamentally as actively participating beings, and we know the world and its properties solely through practice. Theoretical reflection is derived from more fundamental practical actions within the world. Why does this represent a fundamental shift in perception? How had people previously regarded humanity's ideas? In order to understand this, it is helpful to briefly revisit some of the philosophies that were introduced previously in this book to draw a contrast with pragmatism. These are competing visions regarding the nature of ideas:

Platonism: As discussed previously, Plato's metaphysical philosophy forms the basis for much of later Western philosophy. Plato understood ideas as eternal and unchangeable entities representing that which is truly real behind the appearances with which people interact in the course of their daily lives. For Plato, ideas represent reality's basic roots, so to speak. Ideas are eternal, substantial, and static. Ideas are things we discover and uncover in our exploration of the world. There is, moreover, a cosmic hierarchy of ideas in that the primary idea that empowers all other ideas is the idea of the good, as illustrated in Plato's influential allegory of the cave, in which the idea of the good is compared with the sun, which sheds light on all other beings (Plato, 1987). The pragmatists lived in a world in which Darwin's theory of evolution seemed more plausible than Plato's doctrine of ideas, and they could thus not accept an unchangeable, idea-centered world concealed behind nature's changeability. For pragmatists, a horse is not a horse because it partakes of the idea of a horse; it is a horse because its ancestors were selected on the basis of environmental factors, leading to gradual adaptation.

Cartesianism: We saw previously how modern philosophy began with Descartes. Whereas Plato's philosophy and Ancient and Medieval philosophy in general were a product of metaphysics (i.e., of a theory concerning the essential being of all things), philosophy shifted focus with Descartes and became centered on epistemology. The fundamental question was no longer "How is the world fundamentally constituted?" but, rather, "How can I know the world?" Galileo and later Newton replaced the classical teleological worldview of Plato and Aristotle with a mechanical worldview. In the words of the sociologist Max Weber, the world was disenchanted. In the disenchanted world, ideas are no longer regarded as objective, as inherent in the world; they are regarded as constructions made by the minds of people on the basis of subjective experience. For Descartes, ideas were the name for consciousness' subjective contents. As I demonstrated previously, this train of thought was taken up by British empiricists such as Locke, Berkeley, and Hume. A sort of dislocation took place in the history of ideas, with ideas moving from being seen as structuring principles in the world's design (as in Plato) to being humanity's representation of the outer (disenchanted) world. The pragmatists would concur with Cartesianism that ideas are things we construct and not things we discover out in the world. They would, however, disagree with the subjectivist aspect of Cartesianism, which argues that ideas are necessarily individual, internal, and inside our minds. Ideas are no more private than are hammers and nails. Ideas possess meaning and significance—and meaning and significance are social and public. Dewey (1925) goes so far as to say that it is "heresy to conceive meanings to be private, a property of ghostly psychic existences" (p. 189). Pragmatists regard human knowledge and ideas as fundamentally social inasmuch as we think and act in social contexts. In Dewey's perspective, people "turn outward" and know the world through active action. It is wrong to believe, like Descartes, that we "turn inward" and know the world through passive observation and construction of ideas within our subjective mind. For Dewey, the rightness of ideas rests within "the cumulative objective appliances and arts of the community, not in anything found in 'consciousness' itself or within the organism" (p. 347). An idea or representation is not true because it is possible for me to subjectively compare it with "the world as it is" but because we benefit from it in our practices as a community.

Marxism: Marxism represents a third perspective on ideas, and one that has more in common with the pragmatic conceptualization. Karl Marx's view of the relationship between ideas and reality (the ideal and the real) changed over the course of his life, and later interpreters of Marx's philosophy have been split on how to understand this relationship. It is worth noting, however, that pragmatism concurs with Marxism that knowledge is a result of our relationship with the world, of our practical activities by which we (collectively) act in the world. Pragmatism would also agree with Marxism that people are not just animals living within a preset history and must be understood on the basis of their historical situation; rather, people are also history-creating animals. What the pragmatic philosophers would be quick to reject, however, is the insistence of some Marxists that we can formulate fixed laws for society's historical development. Dewey rejected the assertion that social and psychological "laws" could resemble those of the natural sciences in this way. As Dewey (1927/1946) argued, "Increased knowledge of human nature would directly and in unpredictable ways modify the workings of human nature" (p. 197). When humans develop ideas about social and psychological phenomena, these ideas may become part of humanity's own toolbox for mastering the world and thereby contribute to changing the reality that they concern. This phenomenon was referred to as a triple hermeneutics in Chapter 4, and it can also be called the problem of reflexivity (Brinkmann, 2004). It is rooted in the fact that knowledge from the human sciences and knowledge from the social sciences simultaneously arise from and alter the fields with which they are concerned. When people develop knowledge about themselves and their situations, they unavoidably change themselves and their situations (Ian Hacking (1995) referred to this as a looping effect). Because of this, pragmatists such as Dewey would insist that our history and society are far less predictable than Marxist thought assumes, and it is thus impossible to possess a strict, natural sciences-type of knowledge of history and society. The pragmatists would also reject historical materialism's stance that ideas (about morality, law, aesthetics, religion, etc.) can be thoroughly understood in terms of how the fundamental economic organization of production operates within society. They depicted the human being as much more creative and unpredictable than did (at least some) Marxists.

According to Dewey, the whole dichotomy between the real (the fundamental workings of society) and the ideal (our ideas about the world) arose in Ancient Greece, where it was associated with the difference between the working class and the ruling (and philosophically contemplative) class (Dewey, 1925, p. 124). This social difference was converted into a metaphysical difference between the material (including tools and means), on the one hand, and the ideal or spiritual (including value and ends), on the other hand. Dewey strongly opposed this dualist ideology, which regards values, meanings, and goals as only conceivable by the contemplative and reflective class (the philosophers in Plato's philosophy) while the class responsible for processing nature is relegated to mindless operations. This dichotomy allowed philosophers (and later, scientists) to reject the possibility that ordinary qualitative human experience could be a source of knowledge. Only knowledge that is achieved by means of scientific methods, such as in laboratory situations, is true. This dichotomy also reflects Western philosophy's multitude of dualisms—means and ends, thoughts and actions, facts and values, subjects and objects, bodies and souls. Pragmatism represents an attempt to show that each of these dualisms is false and unfruitful.

If we return to Menand's characterization of pragmatism as a single revolutionary idea about ideas, then the assertion is that ideas, besides being understood as *tools* in pragmatic thought, must also be understood as deeply *social*, dependent on their contexts and environments, and obliged to adapt as the world develops. Ideas are problem-solving tools relative to concrete and situational problems. This means that an idea's survival depends on its adaptability, not its unchangeability (Menand, 2002, p. xii). If we stubbornly hold onto ideas that no longer solve problems for us, and which perhaps even cause problems, then ideas degenerate from being problem-solving tools to being ideologies that serve no other purpose than upholding the status quo in perhaps oppressing ways. For pragmatists, studying such forms of oppression can be a laudable goal for qualitative social scientists in itself.

Pragmatism's Thinkers

According to William James, a leading figure in pragmatism, this philosophical school was first created by Charles Sanders Peirce in an 1878 article titled "How to Make Our Ideas Clear" (James, 1907/1981, p. 26). Peirce's aim was to show that the significance of ideas and thoughts is determined by the actions they prompt. What, for instance, does it signify if I have the idea that diamonds are hard? It has meaning only in the sense that I will discover that it is very difficult to scratch them. "Hardness" is what hard things do; it is not something that can be passively intuited by contemplating its phenomenological essence. Our concept of an object is, according to Peirce, an understanding of the object's practical consequences and effects. In more simplified terms, things are what they do or are what we can do by using them. This applies not just to tangible objects but also to theories and ideas. The pragmatic method of philosophical thinking thus involves, in James' words, trying "to interpret each notion by tracing its respective practical consequences" (p. 26). When two philosophical theories compete for the truth, for example, we must ask: How would the world be different if the one were true rather than the other? If nothing would be different, then the dichotomy is meaningless, without significance. As James states, there cannot *be* a difference that does not *make* a difference (p. 27).

On the basis of this perspective, James (1907/1981) develops an overarching instrumentalist theory of truth. Truth does not represent a magical relationship between a proposition and the world. Truth is not a characteristic of our utterances, something that they either do or do not possess. When we assess a proposition's truth, James argues that we must test it by posing pragmatism's usual question:

> "Grant an idea or belief to be true" it says, "what concrete difference will its being true make in any one's actual life? How will the truth be realized? What experiences will be different from those, which would obtain if the belief were

false? What, in short, is the truth's cash-value in experiential terms?" (p. 92)

True ideas are those we can verify through practice—that is, ideas that work for us and have "cash-value." As a result, truth is not a static and substantial property of ideas but is something that "happens" to our ideas, as James states (p. 92). Truth is something that is created through action and experience, just like health, wealth, and strength (p. 98). Truth becomes a habit, as some of our ideas adapt themselves over the course of our activities (p. 100). This is, perhaps, the central argument of pragmatic philosophy—and it remains its most contested one. The concept of habits plays an important role among pragmatic philosophers, although in varying ways. The concept represents the core of James' theory of truth, of Dewey's psychology, and of Peirce's ontology. Peirce argued that nature itself was habitual in the sense that natural laws are not transcendental laws that work behind the scenes to control natural processes; instead, natural laws are inherent in and habitual to nature. For Peirce, natural laws have a history and can develop. He thus extended Darwinism to ontology: It is not just organisms and species that adapt to the demands of their environment; the laws of nature do too. The laws of nature are similarly the result of selection and adaptation (Menand, 2002, p. 277).

Peirce is perhaps the most difficult to understand of the pragmatists, but he may also be the most original one. His personal life was problematic, and unlike James and Dewey, he never achieved academic success in his lifetime. Today, however, there is a strong interest in his philosophy, especially in its laying of the groundwork for semiotics—that is, for the scientific study of signs. According to Peirce, signs are inherent in nature and are not linked exclusively with language. His semiotic understanding has found some favor within modern biology among other disciplines. It is also worth noting that over the course of his life, Peirce distanced himself from the pragmatism developed by James and Dewey (Peirce was far too much a classical realist and far too little an "instrumentalist" to accept their propositions), eventually calling his own philosophy "pragmaticism," which he considered to be such an ugly word that no one would want to steal it (Menand, 2002, p. 351).

Dewey's pragmatic theory of inquiry is particularly interesting for qualitative researchers today because it blurs any hard and fast distinctions between scientific knowing and human knowing in general. In other words, for Dewey, science is just a condensed form of human knowing, or a focused form of the activity of coping with the world that we are constantly engaged in as living human beings. In this sense, Dewey's pragmatism is a philosophical anthropology of the human knower. Dewey tried from the beginning of his career to overcome the view of the knower as a passive spectator that we have inherited from the Greeks. He argued that stimuli do not passively impinge on the human senses but instead arise when active knowers are engaged in various activities. This is clear in the following early quote, in which Dewey (1896) discusses the stimulus of a noise:

> If one is reading a book, if one is hunting, if one is watching in a dark place on a lonely night, if one is performing a chemical experiment, in each case, the noise has a very different psychical value; it is a different experience. (p. 361)

This simple example should alert us to the idea that stimuli are constituted only on the background of activities and practices. Experiences are not passive happenings but, rather, functions of human beings' everyday doings and engagements with the world and each other. Contrary to the epistemological tradition from Descartes and the British empiricists, according to which knowledge is a passive representation of the world, it means that there are no experiential elements that are simply *given* in the mind of a spectator. Dewey (1929/1960) wants to replace the image of something being given with the image of something being *taken*:

> The history of the theory of knowledge or epistemology would have been very different if instead of the word "data" or "givens," it had happened to start with calling the qualities in question "takens" . . . as data they are selected from this total original subject-matter which gives the impetus to knowing; they are discriminated for a purpose: that, namely, of affording signs or evidence to define and locate a problem, and thus give a clew [sic] to its resolution. (p. 178)

We take bits and pieces of the world—such as qualitative materials—when we need to do something to go on living in fruitful ways. We discriminate for a purpose, as Dewey says, when we are in a situation with a need to solve a problem. Dewey treated thinking as what we do in such situations when we reflect intelligently (Dewey, 1910/1991). He defined thinking as "active, persistent, and careful consideration of any belief or supposed form of knowledge in the light of the grounds that support it, and the further conclusions to which it tends" (p. 6). Thinking is thus partly about testing the grounds for belief and partly about developing one's beliefs in light of further beliefs that one entertains. In line with fellow pragmatists Peirce and James, Dewey thought that such processes are necessitated in and by a problematic situation, or what he called "a forked-road situation" (p. 11), where one's activity is stuck. All thinking, all forms of reflective understanding of the world, emerge from such situations (more on this later). As an existential affair in this sense, thinking is not something that can be formalized methodologically but, rather, is a whole, active, and situational life process full of emotionality in addition to rationality (Morgan, 2014). Being alive means doing inquiry—inquiry that is contextually bound and therefore calls for qualitative understandings of complex wholes.

The contribution of the classical pragmatists (James, Peirce, and Dewey) can be summarized as follows:

1. Knowing is a human activity of coping with the world. Knowing is related to action—participation in social practices—as the relationship between what we *do* and what subsequently *happens* (knowledge is not a representation of the world, as in the representationalist or spectator theories of knowledge).
2. Knowing is something we do in our everyday lives as much as it is something that takes place in universities and laboratories. In fact, we should only count something as knowledge if it makes a difference to how we experience the world or act in social practices (Brinkmann, 2012).
3. Theories are therefore tools that demonstrate their validity in practice, as aids in problem-solving.
4. The process of inquiry is a general life process that succeeds when an indeterminate situation is transformed into a more determinate situation that can lead us to further and enriched forms of human experience. Qualitative research is simply a condensed and reflective form of such inquiry.

Pragmatism Today

Pragmatism's theory of truth—truth as something useful or expedient—was, and still is, subject to attack. The theory represents a break with the dominant idea in the West that truth is a given and that humanity's task is to discover it, not to invent it. Richard Rorty is an example of a recent pragmatic philosopher who celebrated James' original pragmatic theory of truth (Rorty, 1982). Rorty presented the theory in the context of linguistic philosophy, and his basic thought was that, inasmuch as truth is a property of propositions, and inasmuch as propositions are created rather than found, then truth is created by people rather than found in the world (Rorty, 1989). Rorty said explicitly that he was promoting Dewey's perspective, yet other prominent present-day pragmatic philosophers have criticized Rorty for his presentation of Dewey and pragmatism in general, including his preoccupation with the linguistic side of experience. According to another so-called neopragmatist, Hilary Putnam, what is interesting in pragmatic philosophy is not what Putnam called its clichéd theory of truth, according to which the true is the useful. For Putnum, this turns pragmatism into just another metaphysical theory (Putnam, 1994, p. 152). Instead, Putnam presents pragmatism as a set of theses, which I reproduce here as a way of summarizing the legacy of pragmatism for today's philosophy and social science (I borrow from Brinkmann, 2013a):

> *Antiskepticism*: Putnam finds that pragmatism distinguishes itself with its argument that assertions of doubt and skepticism require justification just as much as do assertions of knowledge. Pragmatism is thus antiskeptical inasmuch as it claims that skepticism must be justified in practice. Possible though it may be to sit around at a university, or like Descartes in his study, and formulate skeptical thoughts (e.g., "How can I be certain that the world exists?"), our ordinary practical and embodied interaction with the world is built on a certainty that we presumably could never match in matters of theory.
>
> *Fallibilism*: Pragmatism rejects metaphysical guarantees for the truth of our convictions. For Putnam, pragmatism is unique as a philosophy in its asserting the possibility—indeed, the

necessity—of being simultaneously antiskeptical and fallibilist. Fallibilism involves an awareness of the perpetual possibility that one has been mistaken and will later need to revise one's convictions.

No fundamental dichotomy between facts and values: According to the pragmatists, normative discourse represents an unavoidable part of scientific, social, and personal life. For Dewey, all sciences are therefore moral sciences because they must all be understood in relation to our practical lives, where they have certain value-oriented preconditions and effects. The sciences "enable us to understand the conditions and agencies through which man lives. . . . Moral science is not something with a separate province" (Dewey, 1922/1930, p. 296). As Putnam (2002) has insisted, every fact loads a value, and every value loads a fact, so it is never impossible to escape normativity, and even the positivists' dream of complete objectivity and validity rests on values (*in casu* objectivity and validity).

The primacy of practice: The primacy of practice in philosophy, science, ethics, pedagogy, and so on may be regarded as the central thesis, which underlies the whole of pragmatism. In everyday language, this simply means that what we do is more fundamental than what we think about what we do. The pragmatists assert that thinking about what we do is itself a practical activity. Understanding the world in terms of practice involves thinking temporally, contextually, and in terms of processes.

All in all, pragmatism can be said to have emerged from the intellectual activities of Peirce, James, and Dewey and those of modern neopragmatists such as Rorty and Putnam. In the 20th century, philosophy became ever more characterized by a divide between British-style analytic philosophy—with an emphasis on clarity, strictness, and scientificity—and Continental philosophy (broadly speaking)—with an emphasis on people's existence as experiencing and interpreting beings. Pragmatism is important as a philosophical discipline because it builds bridges between analytic and Continental philosophy. Or, perhaps more accurately, it regards philosophy's problems in such a way that it transcends the limitations of both of the opposing camps. Like British analytic philosophy, pragmatism looks favorably upon the sciences, but like German and Continental philosophy, it takes a prime interest in the sociological and historical contexts in which the sciences operate.

Pragmatism and Qualitative Research in the Social Sciences

Pragmatism has had quite an effect on later social thought and qualitative inquiry, and here we may think of the Chicago School of Sociology, symbolic interactionism, and also the more postmodern versions that arose later in the 20th century, not least influenced by the work of Rorty (e.g., Rorty, 1982). Already early on, pragmatists were particularly critical of the prevalence of behaviorist science, according to which human beings were seen as mechanically responding to stimuli from the outside (what follows reworks materials from Brinkmann, Jacobsen, & Kristiansen, 2014). Instead, pragmatists proposed that humans are meaning-seeking subjects who communicate through the use of language and constantly engage in reflective interaction with others. According to pragmatic philosophers, human beings are therefore concerned with the situational, the practical, and the problem-solving dimensions of their lives. This is also the case for social scientific endeavors.

In *How We Think*, John Dewey (1910/1991) developed a five-step research strategy or investigation procedure—sometimes also referred to as "abduction" (according to Peirce as a supplement to the approaches of deduction and induction)—according to which the investigator follows five steps toward obtaining knowledge. First, there is the occurrence of an unresolved situational problem, which creates genuine doubt. Second, this is followed by a specification of the problem in which the investigator might also either systematically or more loosely collect data about the problem at hand. Third, the investigator, now equipped with a specification of the problem, by way of his or her creative imagination introduces a hypothesis or a supposition about how to solve the problem. Fourth, the proposed hypothesis is now being elaborated and compared to other possible solutions to the problem, and based on reasoning the investigator carefully considers the possible consequences of the proposed hypothesis. Finally, the hypothesis is put into practice, as it were, through an experimental or empirical testing by which the investigator checks if the intended consequences occur according to expectations and whether the problem is solved or not (Dewey, 1910/1991). This research strategy thus starts out with genuine puzzlement and ideally ends with problem-solving.

In general, pragmatists have therefore been concerned with what philosophers call "practical reasoning" (cf. Aristotle's notion of *phronesis* from Chapter 2). They are thus preoccupied with knowledge that has some practical impact in and on the reality in which it is used. Without privileging any specific part of the methodological toolbox, with its emphasis on abductive procedures, pragmatism has proved very useful, particularly in explorative qualitative research as a framework for problem-oriented investigation. In addition, it has, for instance, inspired researchers working within the grounded theory perspective (Glaser & Strauss, 1967), one of the first self-denoted and systematically described qualitative methodologies, in which the purpose is to create workable scientific knowledge that can be applied to daily life situations.

Pragmatism also heavily influenced the founding of the discipline of sociology on the North American continent. The official "date of birth" of sociology is often regarded as the opening of the first sociology department at the University of Chicago in 1892. The so-called Chicago School of Sociology during the first decades of the 20th century was instrumental in developing the discipline in general, and "members" such as Robert E. Park, Florian Znaniecki, and William I. Thomas were particularly prominent in advancing a specific qualitative stance in sociology. As such, and due to their inspiration from pragmatism, the Chicago sociologists were not keen on theoretical refinement in itself, believing sociology should be an empirical science and not a scholastic endeavor. As Park is known to have said: "We don't give a damn for logic here. We want to know what people do!" Knowing "what people do" thus became a trademark of the Chicago sociologists, and a range of empirical studies from the early 20th century illustrate the prevalence of qualitative approaches and methods such as document analysis, interviews, and participant observation. The Chicago sociologists were keen to get out and study social life directly, often by use of participant observation. The purpose was to create conceptual apparatuses and theoretical ideas based on empirical material.

Inspired by pragmatist notions about the use-value of science, Park wanted sociology through empirical research to be part of public discussions, debates, and politics as a crucial part of modern democratic society (see also Dewey, 1927/1946). According to him, sociologists should leave the library and their offices and go out and "get the seat of their pants dirty in real research," as he once told his students (Park as quoted in Lindner, 1996, p. 81). Moreover, some of the early Chicago sociologists, such as Jane Addams, also pioneered social work and action research and wanted to use sociology to promote social reform. By using the city of Chicago—a city with a population size that increased 10-fold in less than 100 years—as an empirical laboratory for all sorts of investigations, the Chicago sociologists explored—and still explore—city life as a concrete environment for understanding more encompassing social changes and transformations.

Building on the insights from the early Chicago School, several sociologists and social anthropologists, some of whom were themselves students of the early Chicagoans, during the 1940s and onward began to develop the idea of symbolic interactionism, or sometimes more broadly described as interactionism. What began as a distinctly North American project later spread to Europe. Some of the early proponents of symbolic interactionist social science with a strong emphasis on qualitative methods were Charles H. Cooley, Everett C. Hughes, Erving Goffman, Howard S. Becker, Herbert Blumer, and Norman K. Denzin—with Blumer responsible for originally coining the term "symbolic interactionism," which he admitted was a "barbarous neologism" (Blumer, 1969). Symbolic interactionism often refers to the social philosophy of the pragmatist G. H. Mead (who was Dewey's close friend) as the founding perspective, which was later developed, refined, and sociologized by others. Symbolic interactionism is based on an understanding of social life in which human beings are seen as active, creative, and capable of communicating their definitions of situations and meanings to others. According to Blumer, there are three central tenets of symbolic interactionism:

1. Humans act toward things on the basis of the meanings that things have for them.
2. The meaning of such things is derived from, or arises out of, the social interaction that one has with one's fellows.

3. These meanings are handled in, and modified through, an interpretive process used by the person in dealing with the things he or she encounters (p. 2).

As is obvious from this, symbolic interactionists are concerned with how humans create meaning in their everyday lives and with how this meaning is created and carved out through interaction with others and by use of various symbols to communicate meaning. Blumer stated the following on the methodological stance of symbolic interactionism:

> Symbolic interactionism is a down-to-earth approach to the scientific study of human group life and human conduct. Its empirical world is the natural world of such groups and conduct. It lodges its problems in this natural world, conducts its studies in it, and derives its interpretations from such naturalistic studies. If it wishes to study religious cult behavior it will go to actual religious cults and observe them carefully as they carry on their lives. If it wishes to study social movements it will trace carefully the career, the history, and the life experiences of actual movements. If it wishes to study drug use among adolescents it will go to the actual life of adolescents to observe and analyze such use. And similarly with respect to other matters that engage its attention. Its methodological stance, accordingly, is that of direct examination of the empirical social world. (p. 47)

Blumer famously argued for the development of so-called "sensitizing concepts" as opposed to "definitive concepts" to capture social life theoretically. Sensitizing concepts give the user a sense of reference or a way of looking at empirical phenomena. They should not be thought of as eternally true reflections of the social world but, rather, as informing instruments that can assist us in understanding and eventually coping with the different situations and materials we encounter.

Goffman and Garfinkel

One of the main proponents of interactionism was Erving Goffman, who throughout his career, which started at the University of Chicago in the early 1950s, gradually developed a perspective to study the minutiae of social life that still today is one of the most quoted and used within contemporary qualitative social research (Goffman, 1959). Goffman's main preoccupation throughout his career was to tease out the many miniscule and often overlooked rituals, norms, and behavioral expectations of the social situations of face-to-face interaction between people in public and private places, something which at the time was often regarded with widespread skepticism by more rigorously oriented social researchers. Like one of his main sources of inspiration, Georg Simmel, Goffman keenly used the essay as a privileged means of communicating research findings, just as other literary devices such as sarcasm, irony, and metaphors were part and parcel of his methodological toolbox. Goffman was particularly critical of the use of many of the methods prevalent and valorized in sociology at his time (and still today), such as statistical variable analysis and the privilege of quantitative methodology. Instead, Goffman opted for an unmistakable and distinctive qualitative research strategy aimed at charting the contours and contents of the all too ordinary and ever-present, but nevertheless scientifically neglected, events of everyday life.

In his empirical work, Goffman relied heavily on all sorts of empirical material. He conducted interviews with housewives; he explored an island community through in-depth ethnography; he investigated the trials and tribulations of patient life at a psychiatric institution by way of covert participant observation; he performed the role as a dealer in a Las Vegas casino in order to document and tease out the gambling dimensions of human interaction; he listened to, recorded, and analyzed radio programs; and he more or less freely used any kind of qualitative technique, official and unofficial, to get access to the bountiful richness of social life. Building on a general dramaturgical approach to social life, Goffman read human action through different analytical metaphors: "the ritual metaphor" (looking at social life as if it was one big ceremonial event), "the game metaphor" (investigating social life as if it was inhabited by conmen and spies), and "the frame metaphor" (concerned with showing how people always work toward defining and framing social situations in order to make them meaningful and understandable) (Brinkmann et al., 2014). Goffman's perspective later inspired new generations of sociologists in particular, and qualitative researchers in general, who have used him and his original methodology and colorful concepts to study a variety of conventional as well as new empirical domains, such as tourist photography, mobile phone communication, and advertising. One can say that Goffman was intent to "make the obvious obvious" for his readers. But one may also argue that he was a representative of the qualitative strategy of "making the hidden dubious," which is Bjerre's (2015) addition to the three qualitative strategies identified by Noblit and Hare (1988). Critical research wants to make the hidden obvious (e.g., unmasking ideologies and uncovering social structures), phenomenologists want to make the obvious obvious (describing everyday life), but pragmatists typically want to say that basically nothing is hidden behind the sayings and doings of human beings. Persons are the only actors in social life, not social structures or mysterious "social forces." This is a widespread attitude among micro sociologists, and it applies in particular to the work of ethnomethodologists. Like Goffman, ethnomethodologists take an interest in studying and unveiling the most miniscule realm of human interaction, and they rely on the collection of empirical data from a variety of sources in the development of their situationally oriented sociology. Ethnomethodology was initially a project masterminded by American sociologist Harold Garfinkel, who, in *Studies in Ethnomethodology* (Garfinkel, 1967), outlined the concern of ethnomethodology as the study of the "routine actions" and the methods of meaning-making used by people in everyday settings (hence the term ethnomethodology, meaning "folk methods"). These routine activities and the continuously sense-making endeavors were part and parcel of the quotidian domain of everyday life (described by Garfinkel, in the characteristically obscure ethnomethodological terminology, as the "immortal ordinary society") that rest on common-sense knowledge and practical rationality. Garfinkel was not primarily inspired by pragmatism but, rather, by the phenomenological sociology of Alfred Schütz, but his insistence that nothing is hidden in social life—and the corresponding strategy of making the hidden dubious—aligns him with the pragmatists.

Garfinkel concerned himself with the classic question in sociology: How is social order possible? But instead of proposing abstract or philosophical answers to this question, or proposing "normative force" as the main arbiter between people, Garfinkel set out empirically to document what people actually do whenever they encounter each other. True to the general pragmatist and interactionist perspective, ethnomethodologists rely on an image of human actors as knowledgeable individuals who through such activities as "indexicality," the "etcetera principle," and "accounts," in Wittgenstein's terminology, "know how to go on." Social reality and social order are therefore not something static or pre-given behind human action. It is rather the outcome or "accomplishment" of actors' local meaning-making activities amidst sometimes bewildering, confusing, and chaotic situations. As Garfinkel (1967/1984) stated about the purpose and methodological procedures of ethnomethodology, phrased in the characteristically tortuous ethnomethodological wording,

> Ethnomethodological studies analyze everyday activities as members' methods for making those same activities visibly-rational-and-reportable-for-all-practical-purposes, i.e. "accountable," as organizations of commonplace everyday activities. The reflexivity of that phenomenon is a singular feature of practical actions, of practical circumstances, of common sense knowledge of social structures, and of practical sociological reasoning. . . . I use the term "ethnomethodology" to refer to the investigation of the rational properties of indexical expressions and other practical actions as contingent ongoing accomplishments of organized artful practices of everyday life. (pp. vii, 11)

According to ethnomethodologists, there are many different methods available to tease out the situational and emerging order of social life that is based on so-called members' methods for making activities meaningful. Ethnomethodology is, however, predominantly a qualitative tradition that uses typical qualitative methods such as interviews and observation strategies for discovering and documenting what goes on when people encounter everyday life, but they also provoke our ingrained knowledge of what is going on. Thus, one particularly opportune ethnomethodological way to find out what the norms and rules of social life really are and how they work is to break them. Garfinkel invented the so-called "breaching experiments" aimed at provoking a sense of disorder in the otherwise orderly everyday domain so as to see what people do to restore the lost sense of order. Of these breaching experiments or "incongruence procedures," which Garfinkel asked his students to perform, he (1967/1984) wrote,

> Procedurally it is my preference to start with familiar scenes and ask what can be done to make trouble. The operations that one would have to perform in order to multiply the senseless features of perceived environments; to produce and sustain bewilderment, consternation and confusion; to produce the socially structured affects of anxiety, shame, guilt and indignation; and to produce disorganized interaction should tell us something about how the structures of everyday activities are ordinarily and routinely produced and maintained. (pp. 37–38)

Throughout the years, Garfinkel, his colleagues, and students performed a range of interesting studies—of courtroom interaction, jurors' deliberations, doctors' clinical practices, transsexuals' attempts at "passing" in everyday life, piano players' development of skills and style, medical staff's' pronunciation of patients' deaths, police officers' craft of peacekeeping, pilots' conversation in the cockpit, and much else—aimed at finding out how everyday life (and particularly work situations) is "ordinarily and routinely produced and maintained" by using breaching experiments, but also less provocative methods.

Before summing up, I illustrate more concretely how the pragmatist approach can be used in qualitative inquiry. Unlike some of the other paradigms discussed in this book (e.g., phenomenology), it is difficult to specify concrete principles or steps when working with pragmatism. Dewey's (1910/1991) depiction of the process of thinking or inquiry from *How We Think* is likely as close as we can get, and this was referred to previously. I have pointed instead to the contributions of sociological classics such as Goffman and Garfinkel because these scholars worked on the background of a broad pragmatist philosophical orientation, but they really assembled their own toolbox when conducting empirical qualitative inquiry. In a way, this is pragmatism in a nutshell because it illustrates the pragmatists' distrust of general systems and principles and instead highlights the necessity of personal, situated, and contextual thinking. I shall, however, in Box 5.1 refer to a specific application of pragmatism to the field of health psychology.

Box 5.1 **Pragmatism and How to Make the Hidden Dubious**

For pragmatists, there is no mysterious relation between their theories about the world and the world itself. They give up what philosophers call the correspondence theory of truth, and instead they advocate a notion of truth and validity as that which is useful relative to human interests. A scientific theory can no more reflect the world as it is than the blind person's cane. Theories and canes are tools that humans have developed in order to solve problems and get on with living in fruitful ways. Thus, hidden correspondence relations, social structures, or deep layers of cultural life are met with suspicion and rendered dubious by pragmatists. Instead, persons are viewed as the agents of social life, and knowledge is a tool for action. The question then becomes how to understand the link between interests, actions, and consequences. In a paper on the relevance of pragmatism in health psychology, Cornish and Gillespie (2009) discuss this question by drawing a pragmatic distinction between different kinds of knowledge interests:

Knowledge for predetermined outcomes: When one operates within a relatively specific and stable context and knows the desired outcome of one's activities, then randomized controlled trials (RCTs) are a helpful way of producing knowledge. This comes from medicine and has been developed to test the relative efficiency of treatments and therapies over other treatments or placebos. Some positivists claim that this is the gold standard in knowledge production, but pragmatists deny this because it is very often not possible to operate within a specific and stable context. There simply is no absolutely "best" method (p. 803).

Knowledge for taking care of oneself: In the context of health research, it is quite often insufficient to know "what works" in the abstract. "What works" will often not work within given contexts for a host of local reasons, and to serve the interests of laypeople, it is usually better to begin with people's experiences (including of the "what works" schemes) and work with analytic procedures such as grounded theory or narrative analysis.

Knowledge for intervention design: Health practitioners work with "the complex, real-world, everyday practicalities of individuals and communities, where familial, financial, political, cultural and social dimensions are deeply entwined with health behavior and outcomes" (p. 804). In this context, it is necessary to ensure acceptability of service users and stakeholders, and therefore more community-based and even activist work (e.g. "conscientization" as articulated by Paulo Freire) can be helpful.

Knowledge for cultural critique: The pragmatists are interested in developing new ideas, but "what sorts of methods produce ideas for 'what might be' rather than 'what is'?" (p. 805). If RCTs and modern positivism study "what works" and phenomenologists study "what is," then pragmatists can be said to study "what there might be" due to their interest in problem-solving for the sake of the enrichment of future experiences. To develop ideas about this, we must employ imaginative methodologies, perhaps with inspiration from the arts.

Cornish and Gillespie sum up and emphasize that pragmatism represents a pluralist approach to knowledge, which, they argue, is not to be confused with relativism. There are more and less adequate ways of solving problems, and a pragmatist will look to (1) people's everyday experience, (2) public deliberation about problems, and (3) constant critique of the choice of interests being served when choosing which problems to solve—and which kinds of ideas to develop in the toolbox of social and human science.

Summary

In this chapter, I have introduced the philosophy of pragmatism from the founders Peirce, James, and Dewey and its application in the social sciences represented by names such as Mead, Blumer, Goffman, and Garfinkel (although the latter is not normally grouped with the pragmatists). In philosophy, there are disagreements between anti-realist pragmatists such as Rorty and realist pragmatists such as Putnam, but all strands of pragmatism conceive of the human being as an active, participating creature who knows the world through acting in it. Methodologically, the core of pragmatism is *abduction*—that is, developing potentially helpful understandings and explanations in uncertain situations that are tested to determine if the situation becomes more clear and workable. Unlike induction (going from many individual instances to general knowledge) and deduction (testing general hypotheses deduced from existing knowledge), abduction begins with a breakdown in our understanding of something and is oriented toward making the indeterminate more determinate in order to facilitate action (Alvesson & Kärreman, 2011). Abduction says "X is strange and incomprehensible and blocks my action—but if Y is the case, then X is less strange and I may act. Thus, I allow myself to claim Y for the time being." "For the time being" is a crucial addition because the pragmatists do not believe is static and eternal knowledge about the social world. In that sense, they are Aristotelians and understand the social world as changing, partly in response to the knowledge that we develop about it.

I have also argued that the pragmatist research ethos can often be described as "making the hidden dubious" because there is a focus on action—what we do, how we experience it, and what the consequences are—rather than on hidden social structures or deeper layers of the social world. For some micro sociological pragmatists, such constructs are treated as faulty illusions or myths (Harré, 2002) that we must criticize philosophically. It is not always easy to assess whether this pragmatist ethos is a form of realism or the opposite, and there are many differences among people who refer to themselves as pragmatists. But I believe that Morgan's (2014) characterization is apt and places pragmatism between anti-realist constructivism and realism: "On one hand, our experiences in the world are necessarily constrained by the nature of that world; on the other hand, our understanding of the world is inherently limited to our interpretations of our experiences" (p. 1048). Experience, for the pragmatists, however, should not be understood as a subjective phenomenon enclosed inside the skull of an individual but, rather, is a name for the whole process of acting, encountering problems, suffering the consequences, and trying to provide remedies in a whole cyclical movement. This is a dance involving both subjects and objects, knowers and the knowns (Dewey & Bentley, 1949/1960), and it is therefore very difficult to pinpoint using traditional philosophical terminology. Pragmatists reject the whole epistemological setup of knowledge as a representation of a world that is independent of the practice of knowing and instead portray knowledge as "coping skills," developed collectively by engaged human beings. Because of its commitment to real-world problems, pragmatism "has the advantage of naturally assigning a central role of politics and ethics in every aspects of human experience" (Morgan, 2014, p. 1051). This is so because if "knowing is doing," then it is pivotal to consider in every case what doing *well* means. As such, pragmatism is particularly well-equipped to discuss the ethico-political presuppositions and implications of qualitative research practices.

Quite often, pragmatism is discussed in qualitative inquiry in relation to mixed-methods research—that is, research that employs both quantitative and qualitative procedures. Because of their commitment to what is useful in practice, the pragmatists are seen as providing arguments in favor of mixing methods. This may be so, but I agree with Denzin and Lincoln (2011b) that the links between mixed methods and pragmatism are problematic and that advocates for mixing methods do not always understand the philosophical underpinnings of pragmatism (p. 246). For that reason—and because this book is about philosophical ideas of qualitative research specifically—I have omitted the discussion of mixed methods.

6

FRENCH PHILOSOPHIES OF QUALITATIVE RESEARCH

STRUCTURALISM AND POST-STRUCTURALISM

SOME OF THE most important philosophical impulses in the 20th century came from France. In Chapter 4 on phenomenology and hermeneutics, I addressed how the phenomenological ideas of Merleau-Ponty and Sartre gained in prominence, and it seems fair to say that Sartre's existentialism was *the* philosophy of the 1940s and 1950s, appealing not just to philosophical colleagues throughout the world but also to many lay readers of Sartre's books. I should also mention Albert Camus, who is normally considered an existentialist philosopher, even though there are considerable differences between Sartre and him, the former arguing that "existence precedes essence," whereas the latter was more inclined to speak of a human nature that resulted not just from the individual's radical choice but also from being as such. Both Sartre and Camus were famous also for their novels and plays and for notable political engagements.

But by the mid-1960s, existentialism was being seriously challenged as a philosophy in France, especially by structuralism, which presented itself as a more scientific mode of philosophizing (Ingram, 2010). The linguist Saussure had already paved the way by developing a structural theory of language, and especially the structuralist anthropologist Lévi-Strauss exerted much influence across the human and social sciences. Also, Marxism became strong as a philosophical school in the latter half of the 20th century, both in and outside of the academy, with violent student revolts and the birth of various anti-capitalist countercultures. Sartre had already tried to unite existentialism and Marxism in his later writings, and in the structuralist camp it was Althusser who became the most influential Marxist (p. 14).

In this chapter, I give an introduction to the philosophical ideas of structuralism and also try to show how they eventually developed into post-structuralism, which has long been a very significant philosophical orientation for qualitative researchers and remains so today. I focus first on the founder of structuralism, Saussure, and also refer briefly to Lévi-Strauss and Althusser before moving on to the post-structuralists, concentrating on Derrida and Foucault, who were in fact quite different philosophically. I end by discussing how structuralism and post-structuralism have informed qualitative inquiry specifically in the guises of discourse analysis and deconstruction. Following the scheme found in Noblit and Hare (1988)—and the distinction between making the hidden obvious, the obvious obvious, and Bjerre's (2015) addition of making the hidden dubious—we can say that the post-structuralists quite clearly represent a fourth strategy of *making the obvious dubious*, especially in the deconstructionist versions. For the goal is here to take what we otherwise take for granted and show that our understanding of it is contingent and therefore *could* be different, *will* be different (because meanings are unstable and subject to change), and perhaps *ought* to be different for ethical or political reasons. So, post-structuralist philosophy is for those qualitative researchers who like to rock the boat, so to speak, and unsettle the seemingly stable foundations for our knowledge and lives, and show that our constructions of gender, morality, and the subject itself are set up within problematic binaries that qualitative studies can and should trouble.

Structuralism

Structuralism began as a theory about language, developed by the Swiss linguist Ferdinand de Saussure (1857–1913). Philosophers have long wondered how language obtains its meaning. How is it that the sounds we emanate and the signs we scribble down can signify all sorts of marvelous things? According to structuralism, the answer is not to be found by looking at words, sounds, or signs individually but, rather, by taking the whole structure of language into account. In his main work, *Course in General Linguistics*, published posthumously in 1916, Saussure emphasized that "language is a system all of whose terms are interdependent and in which the value of one term arises only from the simultaneous presence of the others" (Saussure as quoted in Broden, 2010, p. 233). Thus, no linguistic unit has meaning in and for itself, but only in relation to other units within an overarching system. The word "father" thus obtains meaning not by direct reference to individual beings in the world but by standing in a relationship to "mother," "child," "grandfather," and countless other elements within a linguistic structure. In this context, Saussure introduced his famous distinction between *langue* and *parole*. The former refers to language as a system, characterized by Saussure as "a self-contained whole and a principle of classification" (p. 234). *Langue* is "never complete in any single individual, but exists perfectly only in the collectivity" (p. 234). It is thus not an individual psychological phenomenon but, rather, a structure that truly exceeds the minds of persons. *Langue* is a "fund accumulated by the members of the community through the practice of *parole*" (p. 234), which means that the latter, *parole*, is language as it is spoken and practiced.

So we see that linguistics became established as a science the moment it was given an autonomous phenomenon (*langue*), detached (or semi-detached, because *langue* and *parole* are indeed related) from other spheres of the world. This also means, however, that meanings become detached from the objects they designate, something the German logician and philosopher of language Gottlob Frege had also articulated through his distinction between meaning and reference (*Sinn* and *Bedeutung* in German). Two words can have the same reference (e.g., the planet Venus) and yet have quite different meanings, such as "the evening star" and "the morning star," respectively (without a speaker necessarily knowing that these two words refer to the same astronomical object). For Frege, this was proof that the meaning of a word is not to be found in the object designated (its reference or *Bedeutung*), and Saussure would point to the system of *langue* as the fund of meaning. For structuralists such as Saussure, meaning is not established with reference to something beyond language as a system but, rather, by looking within the structure, where meaning arises from *difference* within language ("father" is thus defined as *not* "mother," *not* "child," etc.).

Broden (2010) sums up the legacy of Saussure and argues that "structural linguistics is known for having inspired some of the most brilliant minds of the postwar generation in Europe to renovate their field through theoretical reflection and the development of explicit methodologies defined in precise terminologies" (p. 241). After Saussure, many different social scientists would take the structuralist framework and apply it to organizations, cultural practices, politics, psychological disorders, aesthetic objects, and many other things (p. 242). In linguistics, the renowned Noam Chomsky developed his own Saussurian version of a structural theory of language, and many others "fostered projects for qualitative 'human sciences' based on observation, exacting analysis, and interpretation" (pp. 222–223) in a way that provided for a genuine alternative to the dominance of the otherwise behaviorist approaches of most American social science at the time.

The most famous structuralist in anthropology was Claude Lévi-Strauss (1908–2009). His structuralist anthropology sought to go beyond people's experiences and focus on underlying structures in society. Just as there is a difference between *langue* and *parole* in language, so there is a difference between how people act and experience their lives phenomenologically (equivalent to *parole*) and the way that the structures of kinship, classifications, and myths operate beyond experiences, actions, and practices. By decentering the acting and experiencing human being, Lévi-Strauss' structuralism represented a form of anti-humanism, according to which humans are not only unaware of the underlying structures that order their lives, "but that it is the structures, not the men, doing the thinking" (Singer, 2010, p. 251). The hermeneutic philosopher Paul Ricoeur characterized Lévi-Strauss' structuralism as a "Kantianism without a transcendental subject" (as quoted in Singer, 2010, p. 250), which means that the focus was on the structures that order human experience and action (somewhat equivalent to Kantian categories that construct human experience), but without a human subject in focus that actually suffers the experiences or does the actions.

Louis Althusser's Marxist structuralism was an even more radical anti-humanism. Althusser was an influential historical materialist who argued against all humanistic interpretations of Marx.

He is best known today for his analysis of how a human subject comes into existence through *interpellation*, or a form of "hailing." The famous example given is when a police officer calls a person in the street by saying "Hey you!" The person will then likely turn and face the authority and is in that instance interpellated ideologically by the structure of the system. The person becomes a subject by becoming aware of him- or herself in relation to a state apparatus that functions ideologically. So, also for Althusser, it is the structure that acts through persons to create subjects—and not autonomous persons that act toward each other as in humanist philosophies. This, obviously, is an idea that has attracted much criticism. It also entered French psychoanalysis with Jacques Lacan, who interpreted the Freudian unconscious as being organized as a language. Although the philosophical influence of Althusser has waned, contemporary philosophers maintain an interest in Lacan, represented, for example, in the work of Slavoj Žižek, whose contemporary cultural influence is reminiscent of Sartre's in the heyday of existentialism. Žižek is an influential cultural commentator who interprets various phenomena in the capitalist era in ways that are both funny and thought-provoking, inspired by Hegel, Marx, and Lacan.

Post-structuralists

The literal meaning of post-structuralism is simply "after structuralism." It is a huge and very diverse philosophical tradition that is united more by the legacy from structuralism, which is constantly criticized, than by any positive doctrines. Schwandt (2001) thus defines it as "the general name for a critique of structuralism that arose largely in France in the early 1970s" (p. 203), but adds that it shares many of the same concerns as structuralism, including its hostility toward metaphysics and humanism and also its emphasis on the centrality of language. Post-structuralists generally decenter the subject, the I, and see it as an effect of discourses; they consider all meaningful phenomena as texts that are related to other texts (intertexuality); and they consider meaning as inherently unstable and indeterminable (p. 203). The main analytic strategy in post-structuralism is deconstruction, which is a practice of reading texts against themselves (this is further discussed later). Schrift (2010) notes that the French post-structuralists can all be said to carry on the work of Nietzsche (p. 6). The structuralists drew heavily on Marx (and to some extent Freud, especially in the case of Lacan), but the post-structuralists used Nietzschean ideas in critical discussions, for example, with Kant (in the case of Foucault), Lévi-Strauss (in the case of Derrida), and Hegel (in the case of Deleuze) (p. 6). Nietzsche's perspectivism and his genealogical approach, tracing how ideas have been formed through history, are important for thinkers who like to render dubious what is otherwise taken to be obvious.

Derrida

In the case of Jacques Derrida (1930–2004), there was inspiration not just from Nietzsche but also from Heidegger, and particularly from the *Destruktion* (of Western metaphysics) advocated by the latter. According to Derrida, Western philosophy has been dominated by a "metaphysics of presence," where existence is equated with what is "present-at-hand" in Heidegger's terminology—that is, immediately available for observation and scrutiny. Derrida criticized this privileging of presence over absence (Derrida, 1970). The story of Derrida is illustrative of how provocative post-structuralism has been to the mainstreams of philosophy (and social science in general). In 1992, Derrida was supposed to receive an honorary doctorate at the prestigious Cambridge University, but many philosophers protested because they viewed him as an intellectual charlatan. Eventually, however, he did receive the doctorate (with 336 votes for and 204 against), but it was primarily scholars from literature and the arts who supported it, whereas the philosophers were against. This is symptomatic of the reception of Derrida's ideas and those of post-structuralism more broadly in the philosophical community.

But why this hostility? It is due mainly to the deconstructive approach to language and meaning, crystallized in Derrida's infamous statement that "there is nothing outside the text" (in French: "il n'y a pas de hors-texte"). In fact, a more precise English translation would be "there is no outside-text." The former translation is provocative because it seemingly claims that everything that exists is "text" and that all phenomena—including physical, chemical, and biological ones—are nothing but "text." Such a metaphysical statement would be both absurd and very un-Derridian, but it fueled the culture and science wars, particularly in the United States, in the 1980s. The latter, and correct, translation simply emphasizes, much less radically, that any text (i.e., anything that has meaning) is dependent on other texts for its meaning. Meaning is relational and depends on difference. As discussed previously, the structuralists argued that there are no meaningful units in detachment from the larger whole, but Derrida corrected structuralism by arguing that this larger whole is itself unstable and perspectival. This means that there simply are no fixed meanings; meaning is forever deferred in endless chains of signs that refer to other signs without ever arriving at the true, authentic meaning. Deconstruction as a practice can be seen as a way of reading textual material (this can also be performances, organizations, and other cultural phenomena), showing how meanings float, and possibly involves suggestions and invitations into other ways that meaning may appear. Derrida made clear that deconstruction is not a method that one employs from the outside of a text. Rather, deconstruction is "already at work within a text" (Haddad, 2010, p. 111). According to Critchley (2007), deconstruction is also not a technique but, rather, more like an ethics of reading against the narcissistic images and established truths of our age. This means that deconstruction is not some activity that may one day be completed and finalized. As long as there are meanings, there will be (a need for) deconstruction.

Although Derrida has been depicted as a kind of nihilistic and irresponsible thinker, this is far from the truth. He was committed to social justice and agitated, for example, against the apartheid regime in South Africa. He also argued that justice is indeconstructible, which in a way is a shield against nihilism. But in order to illustrate Derrida's thinking in action, it will perhaps be helpful to include an example of how he engages in analysis, and I have chosen to briefly present his famous deconstructive reading of forgiveness as a phenomenon. From his analysis of forgiveness, we may get a sense of what could be called the ethics of deconstruction (Derrida, 2001). His famous analysis of forgiveness boils down to the following, formulated as an assertion and a question: "There is the unforgiveable. Is this not, in truth, the only thing to forgive?" (p. 32). It is, Derrida seems to affirm, for we can only truly forgive the unforgivable. For if something is forgivable, there is no reason to forgive it: "Forgiveness forgives only the unforgivable"

(p. 32). Forgiveness, like other phenomena analyzed by Derrida (e.g., hospitality), contains a paradox or an *aporia* (a philosophical puzzle): Forgiveness is a real phenomenon in our lives, and yet it is not really possible. It is only possible as an impossibility. Derrida quotes Jankélévitch, who addressed the Holocaust and said that forgiveness died in the death camps, and then Derrida adds: "Yes. Unless it only becomes possible from the moment it appears impossible" (p. 37). Forgiveness is described as a kind of madness: "Pure and unconditional forgiveness, in order to have its own meaning, must have no 'meaning,' no finality, even no intelligibility. It is a madness of the impossible" (p. 45). If one decides to forgive, because it is a meaningful and sensible response, then it is not really forgiveness. If forgiveness is inserted into the common logic of an economic transaction, its special ethical character is ruined: "If I say, 'I forgive you on the condition that, asking forgiveness, you would thus have changed and would no longer be the same,' do I forgive?" (p. 38). No, seems to be Derrida's answer, for forgiveness is unconditional and not an instrument that one may use in order to achieve something. It is, in a sense, its own purpose. But this amounts to a kind of madness in our time that is so preoccupied with the useful and the instrumental.

The deconstructive reading of forgiveness exemplifies how Derrida worked philosophically by letting inner tensions and oppositions of phenomena ("texts") play out. It is not the analyst who, from afar, imposes something on the phenomenon, but the phenomenon itself that deconstructs through its own working. Because of his insistence that deconstruction is not to be thought of as a "method," it has not materialized as a qualitative research technique as such, but the philosophical attitude of deconstruction—of rendering the obvious dubious—has been quite influential among qualitative researchers, especially for what is now known as "post-qualitative researchers," whom we shall encounter in Chapter 7 (Box 6.1).

Foucault

The best known of the post-structuralists, and actually the most cited author from the humanities across all disciplines, is Michel Foucault (1926–1984). Foucault had a mixed disciplinary background and was educated in psychology and philosophy, but he

Box 6.1 **Deconstruction and How to Make the Obvious Dubious**

Even if deconstruction cannot be formalized as a qualitative research method, it has nonetheless inspired many qualitative researchers. In their book on qualitative interviewing, Brinkmann and Kvale (2015) argue that analyses of interviews may borrow from the spirit of deconstruction in the sense of destructing one understanding of a text in order to open for construction of other understandings (Norris, 1987). Qualitative interviewers can focus not on what the person who uses a concept means but, rather, on what the concept says and does not say in a conversational encounter. Deconstruction is here affiliated with a critical "hermeneutics of suspicion," but in line with conversation and discursive analyses (see Box 6.2), it does not search for any underlying genuine or stable meaning hidden beneath a text. Meaning is understood in relation to an infinite network of other words in a language.

As Brinkmann and Kvale (2015) emphasize, deconstructive readings tear a text apart; unsettle the concepts it takes for granted; and concentrate on the tensions and breaks in a text, on what a text purports to say and what it comes to say, as well as what is not said in the text, and on what is excluded by the use of the text's concepts (p. 263). A deconstructive reading reveals the presuppositions and internal hierarchies of a text and lays open the binary oppositions built into modern thought and language, such as true/false, real/unreal, and subjective/objective. Deconstruction does not only decompose a text but also leads to a redescription of the text. A deconstructive analysis could, for example, focus on selected interview excerpts and concepts and address the meanings expressed, as well as meanings concealed and excluded in the text. Steinar Kvale, a pioneer in the development of qualitative research in general and interviewing specifically, even turned to self-critical deconstruction of the ideas that he had spent his career developing (Kvale died in 2008). One of the last pieces he wrote was a deconstructive reading of "the interview dialogue," which appeared in his final book, the updated version of *InterViews* (of which I was co-author), and I can here provide a glimpse of the reading (reworked from Brinkmann & Kvale, 2015):

> We may start by wondering why the two similar terms "interview" and "dialogue" are often added together, and

not uncommonly bolstered with embellishing words, such as authentic, real, genuine, egalitarian, and trusting. Dialogue exists in a binary opposition to monologue, which today may connote an old-fashioned, authoritarian form of communication. When dialogue is used in current interview research it is seldom in the rigorous conceptual Socratic form, but more commonly as warm and caring dialogue. "Dialogical interviewing" can suggest a warm empathetic caring, in contrast to alienated and objectifying forms of social research, such as those found in experiments and questionnaires. When the interview is conceived as a dialogue, the implication is that the researcher and subject are egalitarian partners in a close, mutually beneficial personal relationship. The expression "interview dialogue" here glosses over the asymmetrical power relationships of the interview interaction, where the interviewer initiates and terminates the interview, poses the questions, and usually retains a monopoly of interpreting the meaning of what the interviewer says.

We can further note that the term dialogue is used today in texts from a variety of fields, such as management and education, which advocate "dialogue between managers and workers" and "dialogical education" (Kvale, 2006). In these contexts, with obvious power differences and often conflicts, the term dialogue may provide an impression of equality and harmonious consensus. We will conclude this brief deconstruction of the phrase "interview dialogue" by asking whether the term dialogical interviews used about research interviews, may, corresponding to dialogical management and dialogical education, serve to embellish the power asymmetry and cover up potential conflicts of research interviewers and their interviewees. (p. 263)

This short example of a deconstruction of a central qualitative research idea demonstrates how the obvious (that dialogical forms of interviewing are good and perhaps even emancipatory for the participants) can be made dubious by relating the idea to other meaning-giving contexts, which provides for a different—and less flattering—understanding of the phenomenon.

always attempted to understand his subject matters in their historical contexts. Late in his career, he claimed that his overarching goal had always been "to create a history of the different modes by which, in our culture, human beings are made subjects" (Foucault, 1994, p. 326). I shall take this assertion as a starting point for presenting his work here.

Foucault's historical and philosophical argument involves the view that the human subject as studied by the social sciences did not exist as an "object of truth" until early in the 19th century (Smith, 1997, p. 857), when the social sciences created a subjugated subject by rendering it calculable, manageable, and governable (what follows builds on Brinkmann, 2005a). He is probably most widely known for his critique of modernity as presented in *Discipline and Punish* (Foucault, 1977). Foucault was here intent to examine how modernity's social scientific disciplines historically emerged as attempts to govern and discipline individuals and populations. In modernity, new disciplines, institutions, techniques, and discourses made possible new forms of power exertion. In contrast to earlier, pre-modern forms of power, such as public transportation of prisoners in chains and public executions, modern forms of power work in subtler and more hidden ways, in closed prisons and bureaucratic structures supported by scientific expertise. Foucault's point is not that earlier kinds of visible brutality were morally better than modernity's veiled brutality but, rather, that we must develop new ways of analyzing power in order to account for its modern forms. We can no longer work with a conception of power as centered with a subject, a single sovereign (e.g., a king or a leader), because modernity's power relations are much more diffused and distributed in heterogeneous practices. Modern power is capillary, as Foucault said, no longer connected to obvious physical superiority but, rather, to truth, knowledge, and expertise, dispersed across social life.

A radical point in *Discipline and Punish* (Foucault, 1977) is that the human subject is an effect of power. It is not primarily subjects, who, from a position outside power relations, intentionally exercise power in order to promote their specific interests, for being a subject with interests in the first place is only possible because of power relations. Power is thus not merely oppressive, according to Foucault, but also productive: It produces subjectivities. How should we understand this strange claim? Foucault's illustrative example is the *panopticon*, a prison structure developed by Jeremy Bentham, the father of utilitarian moral philosophy. This structure would enable a single guard to monitor the behaviors of a large number of prisoners. Foucault was interested in this architectonic structure because it represents the ideal type of modernity's rational-bureaucratic form of power exertion, and it well illustrates the point that power should always be understood, not abstractly as belonging to atomistic subjects or different classes, but as concretely embedded in technologies, practices, and physical objects. The panopticon should render individuals visible at all times, but it should also make the individuals monitor themselves. Prisoners unavoidably turn the guards' gaze toward themselves, and thereby become constituted as self-monitoring or self-examining subjects. Here are reminiscences of a theory of self-consciousness as found in the work of pragmatists such as Mead (1934/1962), where human beings always relate to themselves through the gazes and reactions of others. In Foucault, however, this self-relation is portrayed as a relation of power, where subjects are constituted as self-examining individuals. A self-examining individual has the soul, psyche, or subjectivity as a correlate (Coles, 1992). In Foucault's eyes, the soul is the prison of the body. The soul (or mind) is that which is produced when the body is worked upon and disciplined in specific ways. This can be called a physicalistic aspect of Foucault's argument.

Foucault's point is that the self-monitoring subject was constituted in modernity through a number of institutional practices: in schools, factories, prisons, hospitals, courts of law, and the whole system of treatment of individuals at large. The soul, the inner psychological world, became the panopticon that we all carry within ourselves (Coles, 1992). Psychology became the science of the inner world, and therapies and other similar power techniques became normalizing, subjectivity-constituting social technologies, especially when Western men and women became what Foucault called confessing animals, constituted as subjects through confessional technologies (Foucault, 1980). We *are* not simply subjects. Rather, we are *made* subjects through processes of subjectification.

In the earlier works by Foucault, especially *The Order of Things* (Foucault, 1966/2001), he had been working "archeologically" to trace the historical emergence of different *epistemes*, as he called them. These are background conditions that enable specific discourses and forms of knowledge to appear as meaningful in different epochs. The idea of the episteme is quite close to Kuhn's notion of paradigms that we encountered in Chapter 3. When epistemes change, it is not because humans somehow get closer to the truth but, rather, that new possibilities emerge for things and phenomena to be articulated. Foucault wrote about the shift from one episteme to the next: "Not that reason made any progress: It was simply that the mode of being of things, and of the order that divided them up before presenting them to the understanding, was profoundly altered" (p. xxiv). This is how he characterized the shift from the "Classical age" to the "Modern age," beginning approximately 1800. Famously, Foucault ends *The Order of Things* by predicting the end of the episteme in which the human being occupies a central and privileged place. "Man is a recent invention," he wrote, "and one perhaps nearing its end" (p. 422). Following the end of the human-centered episteme, "one can certainly wager that man would be erased, like a face drawn in sand at the edge of the sea" (p. 422). For posthumanists, some of whom are discussed in Chapter 7, this is a kind of prophesy that should pave the way for new non-humanistic ontologies and new materialisms. The question is: does "the death of man," which, of course, is a much disputed claim, also lead to "the death of qualitative research," if the latter is intimately connected to a kind of humanism and emphasis on a knowing, experiencing, and acting human subject? This is a question I address in Chapter 7.

In the final books and articles written by Foucault before his untimely death in 1984 (he was an early victim of the AIDS epidemic), his interests shifted from external power techniques to what he called technologies for individual domination (Foucault, 1988a). He became interested in "the history of how an individual acts upon himself, in the technology of self" (p. 19). Technologies of the self are tools with which an individual acts on herself to create, re-create, and cultivate herself as a subject. Through his historical investigations, Foucault succeeded in locating different kinds of technologies of the self: the Stoic practice of letter writing, Augustine's confessions, examinations, asceticism, and interpretations of dreams. Contra historical materialists, Foucault claimed that technologies of the self are relatively independent of their socioeconomic and political conditions. For that reason, humans are much freer than they think (Foucault, 1988b). The consequence of the fact that the self is not given us to discover in any universal way is that we must create the self as a work of art (Foucault, 1984). The later Foucault still conceived of subjectivity as an effect of power relations produced in concrete practices, but individuals were now understood as being able to determine, to some extent, which practices they will allow themselves to be constituted by. We can use the productive nature of power to shape our selves, which is what Foucault called freedom. Although our lives are unavoidably lived within what he had called epistemes in his earlier work, we should understand these necessary frameworks as preconditions for human freedom, even if they also have the potential to become oppressive.

Technologies of the self need not be oppressive, but can be linked with ethics as practices of freedom. By "ethics," Foucault meant not the abstract philosophical discipline, but a practice, namely the practice of subjective self-formation (Foucault, 1984). Ethics is the practical relationship a self ought to have to itself, and, in this sense, it can be studied independently of moral codes and also independently of scientific truth. There is no connection at all, according to Foucault, between ethical problems and scientific knowledge. This is his axiological starting point. We can just as little demand that ethics should be true scientifically as we can demand that a work of art should be based on scientific criteria. That is why we must create ourselves as a work of art, according to Foucault. There is no deep, inner truth about human nature or about individual subjectivity to be unearthed by science, so the goal of practicing technologies of the self should not be to decipher such an illusory truth. If this happens, then technologies of the self become oppressive. Then they come to belong in the hands of scientific experts—"masters of truth"—and then such technologies degenerate into techniques of domination. That is why Foucault wants to replace the classical Western ideal *know thyself* with what is for him a more primary demand: *Care of the self* (Foucault, 1988a, 1990). There is no deep truth to be known about the subject, but there is the task of cultivating oneself as a work of art.

Postmodernism

Many other philosophers in addition to Derrida and Foucault are associated with post-structuralism, especially if we expand the concept to include what is sometimes referred to as postmodernism. The latter is a broader notion that is not just philosophical but also concerns cultural movements in general, particularly within architecture, literature, and pop culture. Postmodernism emerged as a concept in the late 1970s, especially with Lyotard's (1984) classic *The Postmodern Condition*. Inspired partly by Wittgenstein's idea of language games—different practical uses to which language is put—Lyotard argued that linguistic practices are heterogeneous and that truth is relative to specific language games. Lyotard defined postmodernism as incredulity toward the grand metanarratives of modernity, such as progress through science or emancipation of the working class. Postmodern thinkers generally give up the belief that history has a certain logic that will unfold and liberate people, and instead of big, universal truths, they advocate small, local truths from specific perspectives and connected to the ability to perform actions, as in pragmatism (Kvale, 1992). Lyotard also diagnosed certain problematic tendencies related to this, particularly what he called "performativity" or the idea that after the grand narratives, science and knowledge can only legitimatize themselves with reference to quite narrow instrumentalist purposes, where knowledge is becoming the central force in capitalist production.

Other postmodern/-structuralist thinkers include Baudrillard, who wrote about the tendency of signs to become detached from what they originally signified and instead simply refer to other signs. This creates a medialized hyperreality—a world of simulacra where we can never reach anything "real" and authentic and where everything in a sense is a copy. The cross, for example, used to be an instrument of torture and execution in Roman times, and with the crucifixion of Jesus, it became a symbol of Christianity. When Madonna wore the cross around her neck in the 1980s and sang "Like a Virgin" and "Like a Prayer," it came to signify a certain sexualization, and when young girls would later wear the cross, it would refer to Madonna.

Postmodernism became a very popular perspective or attitude through the 1990s, but today the term is less rarely heard. Reality is making a comeback, so to say, not least with the new materialisms and ontologies that I address in Chapter 7 and which include material, bodily, and affective dimensions in qualitative analyses rather than primarily symbolic and socially constructed ones. However, the legacy from Derrida of "making the obvious dubious" lives on in various deconstructive endeavors, as does Foucault's notion of the subject as an effect of discourses and power relations. Next, I discuss how the post-structural ideas have more concretely influenced qualitative research practices.

Structuralism and Post-structuralism as Qualitative Research Practices

A large number of qualitative researchers today work with tools developed by structuralists and especially post-structuralists. Foucault's ideas have been particularly influential, and Arribas-Ayllon and Walkerdine (2008) have highlighted three broad dimensions that are important in Foucauldian qualitative analyses of discursive practice (p. 91):

1. The analysis of discourse (i.e., normative patterns of sayings and doings) involves *historical* inquiry: This is what Foucault referred to as genealogy or the attempt to trace how the current beliefs and practices came to be evident.
2. The analysis should attend to *power* in the sense articulated previously: Power is what enables things to happen and persons to act and become subjects.
3. The analysis should focus on *subjectification*—that is how subjects are made and molded in practices, given both discourses about the desired and the forbidden and power relations that enable action.

These broad dimensions are relevant for most discourse analysts, who differ, however, concerning the extent to which they are willing to move beyond what is manifest in the empirical materials to be analyzed. A Foucauldian discourse analyst will be permitted to include background factors about power and knowledge when analyzing the manifest contents of textual materials, whereas researchers, who are more inspired by the approach to discourse found in conversation analysis (Potter & Hepburn, 2008), will be more reluctant to go beyond the text itself. An example of the latter, a conversation analytic approach to a qualitative interview, is provided in Box 6.2 to give an impression of how qualitative researchers can work with discursive perspectives (adapted from Brinkmann, 2013b).

Box 6.2 shows how an empirical analysis based on poststructural discourse analysis can stay very close to what was in fact said in an interview. Others, like Ian Parker (2005), see this as a problematic limitation. Parker has argued that it leads to a "textual empiricism," where one is not allowed to say anything about the phenomenon that is not directly visible in the transcription, and this is of little value in most qualitative research projects.

Foucauldian discourse analysts will be interested in studying how some bit of text emerged as meaningful (what are its background conditions?) and not only in analyzing what the bit means on its own terms. Again, Arribas-Ayllon and Walkerdine (2008) may help us because they have articulated some methodological guidelines for conducting Foucauldian discourse analysis. Using the psychiatric diagnosis attention deficit hyperactivity disorder (ADHD) as an example, these guidelines are as follows (based on Arribas-Ayllon and Walkerdine, 2008, pp. 98–99):

Selecting a corpus of statements: Find a suitable sample of discourses that enables one to throw light on the relationship between rules or norms and what is actually stated. The sample can be based on examples of one's "discursive object" (e.g., what people write or say about ADHD). It can also concern the conditions of possibility for the discursive object (e.g., diagnostic manuals in which ADHD figures) and contemporary and historical variability of statements (how has the diagnosis developed and what are the different stakeholders today). The corpus can then be approached with the following four dimensions in mind.

Problematizations: How are the discursive objects made problematic and by whom? (Who says that ADHD is a problem and not simply a way of being human?)

Box 6.2 **Discourse Analysis in Practice**

....

Edwards' (2004) study is an illustration of how to work with discourse theoretical tools in analyzing qualitative data with particular relevance for psychologists. It begins by outlining the main theoretical premises and "methodological stances" of discursive psychology. These are as follows: First, "discursive psychologists avoid coming to conclusions that analysis can reveal people's true beliefs and attitudes" (p. 32). Instead, they look directly at talk as situated interaction and do not attempt to "see through" what people say to arrive at some truer psychological reality. Second, discursive psychologists focus on "contradictions, inconsistencies, and ambiguities" in human talk and (inter)action (p. 33) as important and interesting. They do not see such features as signs of flawed data. Third, they approach all talk as performative, and they are committed to analyzing any kind of talk in terms of the situated actions it performs, even the most mundane kinds of talk.

Using this methodological stance, Edwards (2004) analyzes a number of extracts from interviews, conducted in New Zealand, where racism appears as a theme. The findings are all informed by discourse analytic theory but are nonetheless rich in concrete empirical detail. The following is one example, which also illustrates how many discursive psychologists prefer to transcribe interviews in a very detailed manner (p. 41):

1. R: It's *norm*ally that- Okay *that* argument gets put in that Maoris never get
2.   the jobs okay but you look. hh when they turn up for an interview
3. I: Yes
4. R: *What's* he wearing *how's* he sitting
5. I: Yeah
6. R: *How's* he talking > ya'know what I mean< an' there's no
7.   *point* in having a receptionist that picks up a phone "Yeah
8.   g'day 'ow are ya" ((strong New Zealand accent))
9. I: Ye:s (0.4) [mm mhm
10. R: [I mean they someone that is- (0.4) that is
11.   gonna put their clients ar ea:se
12. I: Right (.) [Mm mhm
13. R: [You don't wanna *shop* a- a shop assistant who's *smelly*

"R" is the respondent, and "I" is the interviewer. The transcription contains information about how speakers emphasize certain words, how long the pauses are between words, when in the flow of words the next speaker interrupts, and so on (regarding these conventions for transcribing, see Jefferson, 1985). The analysis follows immediately after the small excerpts, which are all analyzed very thoroughly.

In this case (just to give a glimpse of the analysis), Edwards (2004) argues that talk like this is used to sustain "a negative view of a group of people, while guarding against a disarming accusation of prejudice" by bringing off that view as rationally or accidentally arrived at (p. 40). For example, he directs our attention to line 2, where the bit, "you look. hh when they turn up for an interview," appeals not only to what he calls scriptedness ("when they turn up for an interview") but also to an event's experiential basis ("you look") (p. 41). It formulates what anyone—and not just the speaker here—can (allegedly) see about Maoris as a general category. This, according to Edwards' analysis, is one way in which prejudice and racial discourse works. He also refers to a number of other microdiscursive mechanisms that are racializing, if not outright racist (e.g., how "factual claims" about minority groups are offered as only reluctantly arrived at, and how prejudice is performed—through talk—as something forced by the realities of no alternative).

Edwards (2004) also explains why interviews are methodologically relevant in this case (because it is difficult to obtain "naturally occurring talk" about everyday racism, which most discursive psychologists tend to prefer if at all available), and he reflects upon the limitations of the study, stating that there is no analysis of the cultural and historical background to the discourses mobilized by the participants in the interviews.

Technologies: What forms of governmentality and power are related to the discursive object under consideration? (How is ADHD made visible by diagnostic technologies and how is it treated?)

Subject positions: How does the discursive object enable different ways of acting (e.g., as a ADHD subject or a psychiatrist treating someone with ADHD)?

Subjectification: How do subjects "seek to fashion and transform themselves within a moral order" (p. 99)—for example, by using a category such as ADHD? (For an analysis of exactly this, see Brinkmann, 2014.)

One may use these categories as broad "sensitizing concepts" when approaching one's empirical materials in qualitative analysis, and this will often be an aid for the researcher. In recent years, there has been a more focused interest on subjectification and subject positions among qualitative researchers, and an entire methodological approach called positioning theory has emerged (see Brinkmann, 2010, on which the following discussion is based). As a recent variety of discursive psychology, positioning theory is a particularly useful framework with which to understand the norms and "moral orders" of human lives. It was originally developed as a dynamic alternative to more deterministic theories that stress roles and rules in social life (Davies & Harré, 1999). Rather than invoking static roles that determine how individuals act, positioning theory argues that the concept of positioning can bring forth the dynamic and changeable aspects of social life, where discourses are seen as offering subject positions that individuals may take up—or may be forced into. Instead of reified social structures or transcendent rules allegedly governing social life, positioning theory argues that "rules are explicit formulations of the normative order which is immanent in concrete human productions" (p. 33). Humans may act in ways that can often be described by rules, but according to positioning theorists (who follow Wittgenstein on this point), this does not mean that they follow rules in acting. Positioning theory is built on the premise that the human social world is first and foremost a normative moral order. Normativities establish social order, continuities, and connections across space and time—for example, between situations, reasons, and actions. There is, for example, a normative connection between the fact that someone transgresses a moral imperative and the feeling of guilt that such transgression affords (but not necessarily determines, since humans are agents and not passive elements in causal chains). Positioning theory is a toolbox that is helpful in enabling us to understand the ways people do things and the meanings and normativities ascribed to what they do. A particularly clear account of its main tenets is given by Harré and Moghaddam (2003), often summarized as "the positioning triangle" of position, storyline, and acts (see also Brinkmann, 2007):

1. A *position* is defined as "a cluster of rights and duties to perform certain actions with a certain significance as acts, but which also may include prohibitions or denials of access to some of the local repertoire of meaningful acts" (Harré & Moghaddam, 2003, pp. 5–6). In every social context, practice, or situation there exists a realm of positions in which people are located, and such positions are inescapably moral (Harré & van Langenhove, 1999), in the sense of involving "oughts." If I am positioned as the leader of the expedition, I *ought* to guide the participants to the best of my abilities. This is what it means to lead the expedition. Positions consist of rights to do certain things, to act in specific ways, and also of (moral) duties to be taken up and acted upon in specific ways. In virtue of his or her position, a teacher will normally be positioned as someone with a duty to arrange situations that will help pupils or students learn and develop. Learners, conversely and in virtue of their respective positions, can be said to have a right to be taught, and a duty to respect the institution of schooling.
2. Positioning theorists argue that social life displays an order that can be described by norms and established patterns of development, and such patterns have come to be known as *storylines*, typically "expressible in a loose cluster of narrative conventions" (Harré & Moghaddam, 2003, p. 6). Sometimes participants in some social episode disagree on which storyline is unfolding. If a man is opening a door for a woman, the man may interpret the event according to a storyline of gentlemanship and civility, whereas the woman may interpret the event as one involving male chauvinism that positions the woman as weak and in need of male protection. Like other forms of post-structuralism, positioning theory emphasizes the vague, ambiguous, and negotiable nature of social reality.
3. Every socially significant action, including speech, as Harré and Moghaddam (2003) note, must be interpreted as an *act*—that is, not just as an intended action (e.g.,

a handshake) but as an intelligible and meaningful performance (the handshake can be a greeting, a farewell, a seal, etc.). An action is given meaning within social practices and as part of some unfolding narrative, and once it is interpreted within a given social episode, it is subject to norms of correctness. Like our thoughts and behaviors, emotions are subject to normative appraisal in our everyday lives, and they are constantly evaluated as warranted or unwarranted, suitable or unsuitable. Identifying acts, including emotional acts, presupposes that we identify the reasons on which they are based, and this in turn involves an understanding of the social episodes and practices in which reasons exist. That is, in order to identify some emotional phenomenon as anger, fear, or guilt, for example, it is not enough to introspect or consider ongoing bodily perturbations. Such emotions are responses to objects and situations: Anger can result when one's rights have been violated, fear may occur when one has been threatened, and guilt may ensue when one judges oneself to have violated a moral norm. Different situations give us different *reasons* to be angry, frightened, or guilty. This is what positioning theorists are interested in analyzing: How human beings act and understand themselves and each other through the positions that are discursively available to them.

Summary

In this chapter, I have focused on the French traditions of structuralism and post-structuralism. We have seen how the structuralist philosophy of Saussure, the linguist, and Lévi-Strauss, the anthropologist, was built upon—and also criticized—in the second half of the 20th century by Derrida and Foucault, among others. Today, these post-structuralists inspire a large number of philosophically informed qualitative researchers, ranging from discourse analysts to deconstructionists. I have given examples from Foucauldian discourse analysis, conversation analysis, and positioning theory, but many others could be included. What unites the structuralists, the post-structuralists, and many of those qualitative scholars who work on the basis of post-structuralism is a detachment of meaning from the world. Already Saussure argued that meaning is not found in the references of language but, rather, internally in language as a structure. Once meaning lost its ties to the world as such, the road was open for the postmodernists and others who study discourses, representations, and social constructions as symbolic spheres in their own right. With the deconstructionist approach, we see a further willingness to unsettle the signifiers and show how everything could in principle be otherwise. In other words, everything obvious is always already rendered dubious. The postmodern attitude can be defined as a stance of openness to and tolerance of the idea that there are no stable foundations, no secure ways that our understandings are hooked on to the world. Some see this as frightening, whereas others see it as liberating.

Chapter 7 follows this line of thought in some ways, but it also diversifies in new directions. For some of the most recent developments in philosophies of qualitative research have built on a dissatisfaction with the detachment of meaning and world, of the semiotic and the material, which is a core idea of many post thinkers. Chapter 7 takes us after the posts and includes some of the philosophies that now seek to return to matter and reunite meaning and materiality. We shall also discuss some philosophies that cut across the four geographical areas that have organized the previous chapters (Britain, Germany, the United States, and France), such as feminism and indigenous philosophy.

7

GLOBAL INFLUENCES ON QUALITATIVE RESEARCH

NEW PHILOSOPHIES

THE PHILOSOPHICAL PERSPECTIVES in this book have addressed qualitative research as a set of practices developed to study the human being and human lives. But most of them work with a quite general perspective of what it means to be human. With poststructural and deconstructionist approaches as notable exceptions, the knowing human being is depicted in generic terms, and knowledge itself—despite the different philosophical paradigms within which it is conceived—is often seen as something unitary, whether connected to accurate representation, valid interpretation, or a pragmatic ability to act. This chapter discusses some philosophical perspectives that render the generic conceptions of the human being and knowledge problematic. Some of these perspectives follow from the postmodern approaches articulated in Chapter 6, but some can also be seen as relatively independent from these.

There is today a diverse group of scholars who question many of the assumptions about the knowing human being that we have met so far in this book, and in what follows, I highlight three of these. First, I address some of the feminist philosophies that question the dominance of the male perspective in philosophy. Second, I discuss some indigenous philosophies that do not necessarily accept the Western philosophical tradition(s) laid out in previous chapters.

Third, I present some of the "new ontologies" or new materialisms, which even question the very idea of qualitative research as such and urge us to move beyond existing philosophies to something called "post-qualitative research." All of these approaches are internally quite pluralistic, and I apologize in advance for only being able to scratch the surface here and all too briefly present their ways of thinking and working that are very sophisticated and profound.

Feminisms

When considering feminism in the context of qualitative research, it is important to begin by noting that it is far from a single doctrine or principle. Feminist approaches to qualitative research "continue to be highly diversified, contentious, dynamic, and challenging" (Olesen, 2011, p. 129). Of course, feminist philosophy grows out of the long struggle of women for gender equality, but there is not agreement about what this means. It is common to make a distinction between different waves of feminist thinking in culture at large. First-wave feminism in the 19th century focused on achieving equal rights for women, such as the right to vote and own property, just like men. Second-wave feminism then began to fight social and cultural inequality more broadly, including the more or less tacit assumptions about men and women that are built into society. Simone de Beauvoir wrote *The Second Sex* in the late 1940s and argued that men are typically depicted as the generic kind of human being in most contexts, and women are seen in some ways as a deviant kind of human being. Carol Hanisch stated that "The personal is political!" in 1969, which was in line with other attempts in the 1960s and 1970s to break free from older oppressive structures. The slogan articulates the idea that people's own thoughts, their intimate lives and social relationships, are not outside of the political sphere but, rather, deeply infused with ideological notions of what is possible (and impossible) for men and women. Finally, third-wave feminism is associated with postmodern thinking of the late 20th century (see Chapter 6), when scholars began to question the essentialist ideas that were said to be inherent in earlier waves of feminism. Women might have been oppressed by being positioned as "the second sex," yes, but are sex and gender really universally fixed categories with a specific content? Third-wave feminists such as bell hooks (2000) criticize earlier forms of feminism for speaking mainly on behalf of educated White women and neglecting the experiences of racial, ethnic, and sexual minorities. In this way, many third-wave feminists are aligned with the poststructural philosophy that was discussed in Chapter 6, seeking to make established binaries and understandings unstable.

However, as Olesen (2011) makes clear in a chapter on feminism in qualitative research, there is indeed one dominant theme in feminist qualitative inquiry, namely the issue of knowledges: "Whose knowledges? Where and how obtained, by whom, from whom, and for what purposes?" (p. 129). Knowledge cannot be taken for granted, according to the feminist perspective. It is always knowledge *for* someone—that is, it works to further certain perspectives and ignores others. No knowledge is therefore innocent. Olesen gives a short introduction to various kinds of feminism, including postcolonial feminist thought (arguing that Western feminist models are inappropriate for addressing women's concerns in postcolonial sites); globalization and transnational feminism (focusing on the impact of globalization); standpoint research (taking the perspective of the situated woman); post-structural thought (dismantling the notion of essentialized truths about sex and gender); research employing the stance of lesbian women, disabled women, and women of color; and, finally, what she calls "endarkening, decolonizing, indigenizing feminist research" (p. 134). This points ahead to the next theme of indigenous research addressed later.

A wide variety of more concrete qualitative research methods can be employed on the background of feminist philosophies. There is no one-to-one relationship between feminist philosophy and qualitative research practices. Mary Gergen (2008) gives numerous examples of excellent feminist research conducted on the basis of ethnography, ethnomethodology, action research, discourse analysis, narrative research, Q-methodology, and archival and institutional research. Given the broad scope of feminisms, it is unsurprising that literally any kind of research methodology may come in handy, given the specific focus of the researcher. Being a psychologist, Gergen highlights in particular Carol Gilligan's (1982) famous qualitative interview study, *In a Different Voice*, which examined the experiences of women who had made decisions as to whether or not to have abortions. Gilligan had been Kohlberg's research assistant when he developed his moral psychological studies and formulated the theory of a hierarchy of moral competence, stating that the more abstract and reflective abilities people had in their moral deliberations, the more capable they were in moral terms (Kohlberg, 1981). Gilligan revolutionized the field by demonstrating that there is a different moral voice coming from women, which emphasizes care and not only justice, and which values concrete relationships between people rather than abstract impartiality. Few qualitative studies have had a societal impact comparable to Gilligan's interview study, and she has since carried on and in recent years developed a so-called listening guide, focused on multiple listenings to interview transcripts in order to articulate voice, resonance, and relationship in the material (Gilligan, Spencer, Weinberg, & Bertsch, 2003).

Indigenous Philosophies

If women's voices have been subdued in much standard (qualitative) research, then the situation has been even worse for people in colonized sites. The ethical, metaphysical, and cosmological ideas of indigenous peoples in the Americas, Africa, India, and the Asia-Pacific regions have rarely been taken seriously by scholars who subscribe to Western ideas of scientificity. Fortunately, this is now changing, however slowly, perhaps because of the significance of the current discussion among Western philosophers about mind/body and culture/nature dualisms (Latour, 1993). As Maggie MacLure (2016) stated in her keynote address at the 12th International Congress of Qualitative Inquiry in 2016, "Indigenous ontologies never had a nature/culture divide." The keynote introduced the new materialist philosophies that I shall unfold in greater detail later, and it articulated an interesting point of convergence between these philosophies (represented by Barad, Deleuze, Haraway, etc.) and indigenous thought. The idea of a gulf between a sphere of matter called nature and a sphere of meaning called culture is peculiar to Western metaphysics and has deeply influenced our conceptions of science, including qualitative research. There are signs now that this idea is coming to an end, or at least now meets significant challenges, and this ought to pave the way for other kinds of metaphysics, such as those found in many First Nations philosophies.

Any discussion of indigenous philosophy, however, should acknowledge the enormous diversity among indigenous scholars, who come from many different tribes and nations throughout the world. Thus, it is not possible to represent indigenous ideas as coming in various phases or waves, as with feminism, because they are simply too diverse. However, Bishop (2005), who writes from a Maori perspective (the indigenous population of Aotearoa or New Zealand), manages to spell out some Maori concerns about research that seem to have wider and more general significance (p. 112):

Initiation: Standard practice of getting access to a field and its participants has been developed based on Western individualist ideas. Indigenous peoples are often more collectivist in orientation, which should be taken into consideration when beginning a research process.

Benefits: Likewise, the question of who will gain from the research is significant, and for indigenous people, it is vital that they can benefit from the qualitative inquiries being conducted. Research should not be conducted *on* indigenous people and then taken away from them, but preferably from within, based on a desire among participants to develop or preserve knowledge.

Representation: Colonizers have often represented the experiences of indigenous people in simplified ways from the perspective of scientific authority. This has been harmful for many indigenous populations, who have their own traditions of representation, which should be respected.

Legitimacy: In the case of the Maori, and likely also many other indigenous populations, colonizers have proceeded from "a social pathology research approach that has focused on the 'inhability' of Maori culture to cope with human problems" (Bishop, 2005, p. 112). This needs to be remedied, and the alleged cultural superiority of the West should be criticized.

Accountability: Finally, there should be a concern about researchers' accountability. Who controls the research process—in everything from initiations to text constructions? As Bishop (2005) writes, "Traditional research has claimed that all people have an inalienable right to utilize all knowledge and has maintained that research findings be expressed in terms of criteria located within the epistemological framework of traditional research" (p. 112). "Traditional research" here means "Western research," and the point is that the kind of universalism expressed by Western science is not neutral but in fact serves certain colonial interests. Perhaps indigenous people have an interest in developing other criteria that are more local and situated?

Much qualitative research is built on a colonizing discourse, Bishop (2005) argues, which hides the researcher under a veil of neutrality. But neutrality is not necessarily neutral, if one can put it as a paradox, because it means that "the interests, concerns, and power of the researcher to determine the outcome of the research remain hidden in the text" (p. 129). This easily comes to work against the interest of indigenous people because "seemingly race-neutral research methods are actually embedded with the racist (largely unconscious) assumptions of White sociologists and these analyses in turn serve to blame underserved minority communities for their marginalized status" (Tachine, Yellow Bird, & Cabrera, 2016, p. 279).

It is not always easy to translate the Western philosophical idiom of ontology, epistemology, ethics, and so on, which I have employed in this book, into the ideas of indigenous frameworks—or vice versa. Some will even claim that the very enterprise of philosophy is a Western practice, which means that Hindu philosophy, for example, becomes a contradiction in terms (this was claimed by philosophers such as Heidegger and Rorty; see the discussion in Sharma, 2013). However, given the rather broad definition of philosophy employed in this book—as much more than a specific analytic discipline and connected to a way of life as such (Hadot, 1995)—I believe that this is an unnecessarily narrow conception. Furthermore, it seems that most writers on indigenous research are happy to employ the vocabulary of ontology, epistemology, and ethics, although the translation process is difficult and the different worldviews might at times be almost incommensurable. In a special edition of *International Review of Qualitative Research*, devoted to indigenous knowledge as a mode of inquiry, Ahenakew (2016) raises a warning:

> When Indigenous knowledge is recognized by mainstream knowledge production mechanisms, it tends to be presented

through the frames of Western epistemology rather than on its own terms. . . . Western frames cannot comprehend, for example, the way Indigenous knowledge places animals, plants, and landscapes in the active role of teacher. (p. 327)

We here see a cosmology that is very different from a Western perspective. Another group of indigenous researchers follow up and state that "in an indigenous research framework and paradigm, knowledge belongs to the cosmos and the researcher is just the interpreter" (Venable, Sato, Del Duca, & Sage, 2016, p. 345). Research, therefore, "must stem from a sense of *place*" (p. 345). As we have seen in this book, most Western approaches view knowledge as belonging to the knowing subject(s). Knowledge, thus, is an epistemological affair. But according to this perspective on indigenous philosophy, knowledge is ontological, belongs "to the cosmos," and is only brought forth and interpreted by the researcher in the capacity of knower. In one way, this is very different from dominating constructivist approaches to "knowledge production" in the West, but in another way, it seems to be aligned with the Platonic stream of Western thought, in which knowledge is inscribed into cosmos in the form of eternal Ideas, where human beings can recollect these ideas in a process of knowing.

That the human being wants to know is universal, but what counts as knowledge varies throughout the world, and how knowledge is obtained methodologically is quite diverse. In this book, we have seen how the Western philosophical ideas are connected with qualitative research methods such as interviewing and participant observation, but in indigenous communities, other practices of inquiry exist. As Ray (2016) states, "Indigenous women have longstanding tradition of research; however, this goes unacknowledged because Western concepts of research are superimposed, as the norm" (p. 365). Beading—the craft of attaching beads to one another to make objects or decorations—is one example of a practice of inquiry that can be taken up as a qualitative research method and is discussed thoroughly by Ray. As she explains, beading has an intrinsic relationship with storytelling, relations, and aesthetics, and it is a significant practice in many cultures. Another example is "sharing circles" (Tachine et al., 2016). These are similar to the standard Western practice of focus groups, although they come from First Nations traditions: "Sharing circles is an open-structured, conversational style methodology that respects story sharing within a Tribal cultural protocol context" (p. 278). Many other examples could be mentioned, but suffice it here to say that every culture and every tribe throughout the world has developed its own ways of knowing and transmitting knowledge across generations. Although often based on philosophies that are different from those informing the Western philosophical canon, the concrete methods and practices employed are sometimes close to conventional qualitative research methods, and sometimes quite peculiar to the Western observer. Indigenous methods, in short, are frequently so old that they look new (p. 282).

New Ontologies and Post-qualitative Research

Most of the ideas presented in this book—whether critical realist, phenomenological, hermeneutic, pragmatic, or structuralist—operate with the notion of empirical materials (data) that one can systematically analyze in order to obtain some kind of insight or knowledge. Experiencing and acting human beings are observed or questioned by researchers to learn about their lives and viewpoints. The process of knowing in qualitative research thus privileges individual human beings and their experiences rather than philosophical thought about the world. However, what is now called "post-qualitative research," operating with new and non-humanist ontologies, questions this assumption (what follows reworks sections from Brinkmann, 2017). There is now a significant and increasing post-qualitative critique of what the post-qualitative scholars summarize as "conventional qualitative research" (Aghasaleh & St. Pierre, 2014). The critique amounts to a political, ethical, ontological, and methodological questioning of the assumptions and practices of qualitative inquiry as normally practiced (e.g., in the form of interviewing and participant observation and paradigms such as phenomenology, interpretivism, and ethnomethodology).

The main argument of the post-qualitative scholars is that what we have come to know as qualitative research is tied to a modernist humanism that ought to be abandoned because it is both ethically and ontologically problematic. Just like qualitative research as such, which is divided into many different camps and philosophies, however, post-qualitative research is no unified movement.

Sometimes terms such as new materialisms, new empiricisms, or posthumanism are used as synonyms. Across the different designations lies, however, a shared interest in ontology—that is, the study of what there is (Lather, 2016). As Patti Lather and Elizabeth St. Pierre (2013), two of the most central persons in the post-qualitative movement, state, "Rethinking humanist ontology is key in what comes after humanist qualitative methodology" (p. 629). And "what there is" ontologically, according to the post-qualitative researchers, cannot be comprehended by using mechanical models based on Newtonian physics and the scientificity constructed on its basis. Often, the quantum theory of Niels Bohr is invoked as providing a different ontological ground than Newton's mechanical physics (especially by Barad, 2007). "What there is" also cannot be comprehended, according to the post-qualitative perspective, by focusing on the subjective or experiential side of the divide that came into being historically when the natural sciences prompted an epistemological gap between mind and world, the subjective and the objective, and the symbolic and the material, as discussed in Chapter 2. Qualitative research is simply a faulty enterprise, the argument goes, because it concerns itself with just one side of a false dichotomy. It privileges the field of human experience and subjectivity, but our ways of thinking of this field—and the techniques and methods developed to study it—presuppose a metaphysics that was misguided from the very beginning. As Lather and St. Pierre (2013) state,

> If we cease to privilege knowing over being; if we refuse positivist and phenomenological assumptions about the nature of lived experience and the world; if we give up representational and binary logics; if we see language, the human, and the material not as separate entities mixed together but as completely imbricated "on the surface"—if we do all that and the "more" it will open up—will qualitative inquiry as we know it be possible? Perhaps not. (pp. 629–630)

According to these post-qualitative scholars, refusing the dichotomy between the objective (quantitative positivism) and the subjective (qualitative research of lived experience) means that conventional qualitative inquiry is no longer possible.

From the reflections of St. Pierre, Jackson, and Mazzei (2016), we can summarize the main philosophical ideas behind post-qualitative thinking along three lines: First, matter (or nature) is understood as agentic and always changing. In a way, this is an ancient idea in philosophy, dating back to pre-Socratic philosophers of flux such as Heraclitus, who depicted the world as a constant flow of becoming (see Chapter 2). We cannot step into the same river twice, Heraclitus famously said, for the river constantly changes and the subjects entering it are likewise in perpetual flux. Matter is not simply cold and dead, to be studied by mechanical sciences, but a warm and vibrant process that acts and develops. This idea was taken up much later by Nietzsche and also by John Dewey, for example, whose pragmatism posited a "metaphysics of events," as articulated in *Experience and Nature* (Dewey, 1925). Even more congenial to the new materialism is Dewey's *Knowing and the Known* (written with Arthur Bentley in the late 1940s; Dewey & Bentley, 1949/1960), which drew explicitly from Bohr's quantum theory, just as Karen Barad has been doing in recent years. More recently, actor–network theorists such as Bruno Latour (2005) have granted agency to non-human "actants" within networks of practices. In his classic, *We Have Never Been Modern*, Latour (1993) argued that the separation of nature (of which we may allegedly obtain objective knowledge) and society/culture/the human (for which we need phenomenology and hermeneutics) was never possible, although this was the backbone to the Enlightenment and the whole modern project, *and* that from which the notion of qualitative research emerged. All in all, this first point deconstructs the constructed opposition between a sphere of passive, inert matter, on the one hand, and a sphere of meaningful human experiences, discourses, and actions, on the other hand. Some follow Haraway (1991) and talk about an "entanglement" of the material and the semiotic—human living, thinking, and acting is always also material, just as the material is always also semiotic. The point of emphasizing the long historical line from Heraclitus up to the contemporary scholars is not to demonstrate that there is nothing new about the post-qualitative movement but, rather, to situate what is new (notably the critique of what they call "conventional qualitative research") in the context of recurring philosophical ideas.

Second, as argued by St. Pierre and colleagues (2016), there is "a heightened curiosity and accompanying experimentation" concerning existence among post-qualitative researchers (p. 102). This means that thinking in the post-qualitative movement goes on not just about "what there is" but more so about what may become (or, to put it in more post-qualitative terms, "what there is" *is* just a process of becoming). As Foucault was fond of pointing out, the goal should not be to discover who we really are, or realize our true, humanist, authentic selves—for these are illusions—but rather to refuse who we are and suggest alternative forms of existence (Foucault, 1988a). Post-qualitative research continues to some extent the playful experimentation that we have come to associate with postmodernism and practices such as investigative poetry, arts-based research, and creative analytical practices in general (Richardson & St. Pierre, 2005), which are also practiced within more theoretically mainstream approaches, but with more emphasis on playing with theory and philosophy. Post-qualitative work begins not with method but, rather, with (posthuman) theory. And theorizing is seen as generative, and there is a real semantic explosion of new words and concepts with the stated ambition of eroding the established binaries that are seen to be inherent in qualitative research. For example, St. Pierre works with what could be called a "flat methodology" (my designation, not hers), which does not erect a hierarchy between so-called data and theory. Normally in qualitative research, we grant a certain privileged position to the empirical data, which are coded, categorized, and analyzed using theoretical concepts, but St. Pierre puts it all on the same plane. Why do we only code what our informants say—and not what Gilles Deleuze or Jacques Derrida say? As Aghasaleh and St. Pierre (2014) spell out, "Knower, language, and the known are all agentive and materially and discursively constructed" (p. 1). This gives us a "flattened ontology" without a hierarchy of something (e.g., "empirical data") being more authentic or closer to reality than something else (e.g., philosophical theory).

Third, and following from the other points, there is a general critique and rejection of the philosophy of representation. The "posts," argues St. Pierre (2011), "announce a radical break with the humanist, modernist, imperialist, representationalist, objectivist, rationalist, epistemological, ontological, and methodological assumptions of Western Enlightenment thought and practice" (p. 615). What Rorty attacked as "the mirror of nature" some decades ago (Rorty, 1980)—that is, the (mistaken) idea that the human mind is a representational device that may mirror, if the proper methods are used, a world that is independent of the mind—is completely dismantled, for there is no detached human being (or metaphorical mirror). This might seem to be a form of social constructionism, but that is a mistaken interpretation, according to the posthumanists, for although social constructionists are aligned with posthumanism concerning their shared critique of essentialism and experientialism, social constructionism lives off the same modernist separation of matter and meaning as conventional qualitative research and simply focuses on the latter (or, in radical versions, claim that there *is* nothing else—the material is also a social construction). Latour for one has been very clear that he dislikes the social constructionist co-option of actor-network theory, given that they (Latour and the constructionists) operate with two very different ontologies. It is for this reason that social constructionist approaches such as discourse and conversation analysis are seen—from the post-qualitative perspective—as belonging to the problematic camp of "conventional qualitative research." These approaches study discourse and symbolic exchanges, but (in most guises) they do not grant agency to matter and thus remain within a problematic form of humanism.

The deconstruction of representationalism is also thought to lead to new kinds of ethics in the post-qualitative movement: St. Pierre et al. (2016) talk about how post-qualitative research implies an "ethico-onto-epistemology, which makes it clear that how we conceive the relation of knowledge and being is a profoundly ethical issue, as is the relation between the human and the non-human" (p. 99). And further:

> If humans have no separate existence, if we are completely entangled with the world, if we are no longer masters of the universe, then we are completely responsible to and for the world and all our relations of becoming with it. (p. 101)

However, there is little in the post-qualitative critique to help us navigate these ethical issues of responsibility, and it also seems logically possible to draw the exact opposite conclusion: If we are no longer masters of the universe, but simply a Deleuzean "fold" or circulating affects, then we have absolutely no ethical responsibility because there is nothing we can control. This discussion is likely to be significant among post-qualitative scholars in the years to come.

Following from the processual and agentic view of nature, the heightened curiosity and experimentation, and also the critique of representationalism—in short, the dismantling of what is called humanism—comes a rejection of "humanist methodology": "The failure of the humanist subject produces the failure of humanist methodology" (St. Pierre, 2011, p. 618). St. Pierre provides a list of conventional qualitative terms that have been deconstructed by post-qualitative scholars, including interview, validity, data, voice, and reflexivity (p. 613). These concepts are said to spring from the notion of the bounded human subjective *self* that should conventionally be called forth through *interviewing*, for example, giving *voice* to the individual and resulting in *data* that should be *coded* and *analyzed* by a separate researcher who needs to engage in a *reflexive process* in order to be clear about her *subjective standpoint*. Something like this, in short, is what post-qualitative researchers frame as conventional qualitative research. And this is what they find deeply problematic.

This can be illustrated with reference to a couple of deconstructive readings of key qualitative terms—for example, coding, which is the practice of assigning labels to bits of one's empirical data in order to enable a quick search in the material and an inductive aggregation of the individual bits into general categories. Coding is conventionally taught as a necessary technique in qualitative research (Gibbs, 2007). For the critics, however, coding reinforces a representationalist epistemology that reduces polyphonic meanings to what can be captured by a single category (see Brinkmann & Kvale, 2015, on which the following discussion is based). According to MacLure (2013), who represents a Deleuzean approach to post-qualitative research, coding offends by positioning the analyst at arm's length from the data. MacLure argues that coding undermines an ethics of responsibility because "researchers code; others get coded" (p. 168). Coding makes everything that falls within its confines "explicable," she says (p. 169), which, from a post-qualitative viewpoint, betrays the ineffable nature of reality. After having delivered a critique of coding from a Deleuzean viewpoint, MacLure interestingly suggests that we can in fact retain it, albeit by understanding it in a different way, namely "as the ongoing construction of a cabinet of curiosities or wunderkammer" (p. 180). Thus, even if there are inherent ontological problems in the practice of coding from a post-qualitative perspective, it may nonetheless be a technique that can spark wonder and creativity in the analyst.

Another term met with suspicion by post-qualitative scholars is the key qualitative notion of experience. We have already seen examples of how St. Pierre and colleagues reject phenomenological approaches in broad terms, and it is notably from phenomenology that qualitative psychologists have learned to place experience on center stage of qualitative studies (see Chapter 4). Critics of the phenomenological studies of experience take issue with the ambition of describing the given, which is criticized as leading to individualist and essentialist lines of work. At the same time as phenomenologists were developing their methods in the 20th century, other philosophers had already been attacking what they saw as "the myth of the given" (Sellars, 1956/1997), arguing that nothing is purely and simply given, and that every understanding is perspectival and rests on theoretical presuppositions. Furthermore, Husserl's original assumption that the goal of phenomenological analysis is to uncover the essences of experiences (see Chapter 4) came to sit uneasily with the anti-essentialist stance of postmodern (and later post-qualitative) thought. Derrida (1970) argued that experience as an idea is connected to what he denounced as a metaphysics of presence. The metaphysics of presence grounds knowledge in what is present to a knowing subject, but, according to "post philosophers" such as Deleuze and Derrida, this is an illusion because there are no stable grounds or foundations from which to know the world once and for all (see Chapter 6). St. Pierre (2008) shows how the qualitative notion of voice (which privileges the speaking subject and her stories) belongs together with "experience" and "narrative" to the questionable metaphysics of presence that we ought to abandon. What the post-qualitative researchers often ignore, however, is the extent to which Derrida's own deconstructive post-structuralism grew out of phenomenology, owing much to Husserl's successor Heidegger, which indicates that there need not be a simple antagonism between phenomenology and its critics.

I am aware that some "conventional qualitative researchers" (in the eyes of the post-qualitative scholars) have also questioned various forms of interviewing, for example, and the whole idiom of "gathering data" and "coding material." From a discursive standpoint, Potter and Hepburn (2005) have criticized the widespread use of interviewing in qualitative psychology, to give just one example, and advocate analyzing naturally occurring talk whenever

Box 7.1 A Top Ten List of Post-qualitative Insights

In her article titled "Top Ten+ List: (Re)Thinking Ontology in (Post) Qualitative Research," Lather (2016) counts down from 10 to 1 and provides the reader with her account of the new materialist philosophy in a way that is short and yet quite helpful. From the observation that

(10) social theory has been intensely language-oriented for quite some time . . .
(9) Lather heralds the recent return to materialism "AFTER Derrida, NOT old school Marxist materialism with its identity politics and economics in the last instance" (p. 125).
(8) She notes that posthumanist theories move away from "the unified, conscious, and rational subject of humanism . . . to the Deleuzean subject" (p. 125), which in general signals . . .
(7) a resistance to the gravitational pull of humanism.
(6) A different canon then gets constructed with Deleuze and Barad as key figures, leading to . . .
(5) new post-qualitative methodologies, and . . .
(4) an affect theoretical orientation concerning researcher subjectivity.
(3) This means that a new theory of social change is called for, namely a kind of Nietzscheanism, which . . .
(2) works by thinking through the body rather than reflexivity as such, and the final lesson is . . .
(1) that "even something as positivist inflected as educational policy analysis can benefit from a neo-materialist approach" (p. 128).

Lather's impressive tour de force through new materialist ontologies and post-qualitative thinking thus ends by connecting the new to existing discourses, signaling that the new materialisms do not so much leave behind the old sets of assumptions as they insert them in a new ontological framework.

possible instead. But the post-qualitative critique is much more radical because it comes from a completely different ontological starting point. This has recently been summarized by Lather (2016) in a useful "top ten list" of insights found in the post-qualitative line of work, summarized in Box 7.1.

Summary

In this chapter, I have tried to move beyond generic philosophies of "the human" and "knowledge," which are otherwise central in the Western philosophical tradition. First, I introduced feminist philosophy, which has challenged the dominance of male-centered perspectives and has also led to many fruitful qualitative studies. Women have challenged their traditional position as "the second sex" and have articulated their experiences as interesting and valuable in their own right. Second, I went beyond the Western colonial approaches and sought to show some of the variety of indigenous philosophies and how these represent qualitative forms of inquiry that may also be inspiring for other (Western) qualitative researchers, not least because of the non-acceptance of the nature/culture divide in most indigenous cosmologies. Finally, I discussed the post-qualitative movement, which also does not accept the nature/culture and matter/meaning divides but, rather, tries to approach practices of inquiry from other theoretical perspectives—for example, inspired by Deleuze, Barad, and Haraway.

I emphasize once again that I have only been able to scratch the surface of these "non-mainstream" philosophies that may inspire and also challenge qualitative research as conventionally practiced. This book is unapologetically mainly about the British, German, American, and French strands of philosophy, but it is hoped that this chapter has served as a reminder that these strands are not agreed upon throughout the world and among all scholars. In the next and final chapter, I summarize, compare the different perspectives and traditions, and try to help the reader decide which philosophy to live by as a researcher and how to incorporate it fruitfully into one's practice of inquiry.

8

DISCUSSION

WE HAVE NOW come to the end of our journey through the times and places of philosophy and its influence on qualitative research. We began in ancient Greece, when Western philosophy was born as a reflection on how reality is constituted and how the human being is placed in the cosmos. Reality was conceived as an *ontic logos*, a well-ordered universe in which value and purpose exist in the things themselves. However, philosophers quickly questioned this, and the perennial problems of realism and anti-realism arose: Are things (or at least aspects of the world) as they are because human beings apprehend them in this way, or is there a definite shape and form to reality that humans may discover? This question is with us to this day and is still discussed by scientists and qualitative researchers.

The idea of a universe with inbuilt value and purpose was shattered after the Renaissance with the birth of modern mechanical science. A whole subject–object split was created: When the outer, objective reality was governed by natural laws, amenable to quantification, the inner, subjective realm had to be born to make room for qualities such as colors and tastes, but also aesthetic and ethical qualities. The ancient Greeks did not have the same kind of subjective–objective dichotomy. But when it arose, it also led to a split between mechanical natural sciences, on the one hand, and interpretative sciences and arts, on the other hand. Philosophers have been divided concerning the question of how to think of the human sciences in this light: Are they like physics and chemistry, as the positivists basically thought, and have the goal of discovering laws of nature? Or are they something else and function by understanding unique human experiences and reasons for action rather than explaining behaviors with references to causes?

Qualitative research was made possible with the split between the subjective and the objective, as it has by and large sought to develop systematic modes of inquiry about everything that does not seem to conform to the practices of inquiry found in the natural sciences. This book has shown how the British traditions of positivism (dating back in particular to David Hume) and realism, the German traditions of phenomenology and hermeneutics, the American tradition of pragmatism, and the French traditions of structuralism and post-structuralism have inspired qualitative research in various ways and provided different kinds of theoretical legitimization for qualitative research practices. In Chapter 7, I discussed some philosophical influences that break with the otherwise neat history told in Chapters 3–6, represented by feminist, indigenous, and new materialist perspectives. The latter, in particular, do not accept the fundamental split between the subjective and the objective, meaning and matter, or the qualitative and the quantitative, which is why some scholars now advocate for "post-qualitative research."

Others have cut the philosophical cake differently and presented other philosophical paradigms in discussions of their relevance for qualitative research. One influential example is the chapter by Lincoln, Lynham, and Guba (2011) on paradigmatic controversies and contradictions in qualitative research. These authors make a distinction between positivism, postpositivism, critical theory, constructivism, and participatory (plus postmodern) approaches. However, I believe that there is a risk that this way of slicing the cake takes the philosophical ideas out of their historical contexts, and one is easily led to false representations of (especially) the older paradigms such as positivism (which is said by Lincoln et al. to be a form of "naive realism," although most positivists were in fact quite unambiguously anti-realist). The authors construct a matrix into which the different approaches are placed, and this is also seen in Lather's (2006) text on "paradigm proliferation," which makes a distinction between positivist, interpretive, critical, post-structural, and neo-positivist philosophies that is somewhat closer to this book's way of representing the different philosophical ideas. Again, however, there is not much historical context, and the different ideas are presented as rather disconnected from what went before and what came after (this is certainly understandable given the short form of Lather's text). In the matrix presented in Table 8.1, I have included the major traditions addressed in this book.

It should now be possible to summarize and compare the major traditions:

Positivism: This is an anti-realist philosophy that eschews the concept of causes that operate beyond what appears to observers. Instead, positivists traditionally work with a phenomenalist and verificationist philosophy, according to which meaningful utterances about the world are true (or false) in virtue of our capacities for testing them scientifically. Knowledge is thus built inductively and must ultimately rest on positive observation. This is achieved through the use of scientific methods that ideally make a sharp distinction between values and pure facts. Methodologically, positivists favor an approach that establishes correlation between variables, or what David Hume called "constant conjunction" between events in experience, which is as close as positivists come to a robust conception of cause. Although many scholars today think of positivism as antithetical to qualitative research, this need not be so, and many influential positivists have historically been open to what we now call qualitative research.

Realism: Realist philosophies come in many different guises, but they are united by their insistence that there is a reality that transcends human experience and that the goal of science is to reach this and uncover its nature. In the social sciences, this perspective has especially been advocated by critical realism with its ontology of layers, arguing that there are "generative mechanisms" in the social world that cause experienced phenomena to appear. These can be studied through qualitative research methods that treat data as evidence for

Table 8.1

Overview of Philosophies of Qualitative Research

	British		German		American	French		Global/Local
	Positivism	Realism	Phenomenology	Hermeneutics	Pragmatism	Structuralism	Post-structuralism	Feminism, Indigenous, and New Materialism
Ontology	Anti-realist, phenomenalism	Hierarchical, layers of reality with basic generative mechanisms	Social world as intersubjective *lifeworld*	Social world constituted by interpretation	Processual reality	Social world as fixed structure	Unstable world, flux	Troubling nature/culture binaries, pluralism
Epistemology	Inductive	Going beyond appearances, seek causes as "inner workings"	Focus on experience	Awareness of prejudices	Knowing through action, abduction	Representationalist, objectivism	Questioning the very premises of the epistemological project	Knowledge is diversified and becomes knowledges
Ethics	Fact/value separation	Moral realism (in some versions)	Moral phenomenology		All inquiry has a moral dimension		Ethics of the other	Communitarian, ethics of care
Methodology	Verify hypotheses, correlation between variables	Seek data as evidence for real phenomena	Descriptive, reduction, *epoche*	Interpretative	Action-oriented, learning by doing	Objectivist	Deconstructive	Rehabilitation of traditional forms of inquiry

real phenomena beyond appearances. Thus, realist perspectives try to make the hidden obvious for researchers.

Phenomenology: Phenomenology is at the same time a basic philosophy of the human being as an intentional creature and a research method that works by bracketing the researcher's foreknowledge in order to arrive at a description of the essence of experienced phenomena. Phenomenology thus tries to make the obvious obvious for us. Some phenomenologists, such as Levinas, have developed sophisticated accounts of ethics based on how we actually experience morality phenomenologically.

Hermeneutics: Hermeneutics is the close cousin of phenomenology, which makes it difficult to place key thinkers such as Heidegger as phenomenologists or hermeneuticists. However, hermeneutics has a clearer emphasis on interpretation in the human understanding of the world, and some—such as Gadamer—would insist that all understanding presupposes prejudices in the form of pre-established knowledge. There is no method as such connected to hermeneutics because understanding cannot be made explicit as a procedure, but it is emphasized that understanding others depends on practical engagement in the world and is not a passive and detached affair. Some have argued that the essence of qualitative research is its basic interpretative or hermeneutic nature.

Pragmatism: A distinct American branch of philosophy, pragmatism claims that knowing is intimately connected to doing, which means that human reasoning is basically practical. Thus, any strict boundaries between facts and values, ends and means, become suspect. Basically, all inquiry has a moral dimension because all inquiry is seen as valid insofar as it results in improvement of the human condition. In some versions (e.g., that of Dewey), pragmatism builds on a process ontology that prefigures those formulated by the recent posthumanist thinkers. I have argued that pragmatism often incorporates the idea that "nothing is hidden" in social life, which means that it often operates with a qualitative strategy of "making the hidden dubious," but it can in practice be coupled with a wide variety of qualitative research approaches.

Structuralism: Structuralism is a body of work that originated in French linguistics and anthropology and became extremely influential in academia throughout the world. Structuralism represents an attempt to map how phenomena of the social world are meaningful because they are related to other elements within larger structural wholes. Studying these wholes aligns qualitative inquiry firmly with scientific practice, which gives structuralist studies a distinct objectivist and representational flavor.

Post-structuralism: Objectivism was criticized by many of the post-structural and postmodern theorists that in one sense followed in the footsteps of the structuralists because of their emphasis on meaning as something arising from relations rather than from reference to the world as such, but that in another sense broke decisively with the scientism of structuralism. Post-structuralists see the world as unstable and meaning as endlessly deferred, and the whole epistemological and humanist project that began with Descartes is called into question. The method (which is not a method in the sense of a process that can be formalized) is deconstruction, which is an attempt to show the precarious nature of meanings. Some post-structuralists understand this as an ethical affair of opening up to other ways of understanding the phenomena of the world by avoiding a reduction of the other to the same. In its deconstructionist guise, post-structuralism embodies the qualitative strategy of making the obvious dubious—troubling what we otherwise take for granted as natural or real.

Global/local: The philosophies brought together under this heading represent a motley bouquet of feminist, indigenous, and neo-materialist approaches that do not accept the male-centered, colonizing, and humanistic discourses that some scholars see as inherent in most other qualitative research approaches. The diverse critiques formulated by representatives of the global/local philosophies cannot easily be homogenized, but they are a vibrant mix of traditional practices of inquiry and radically new philosophies, all of which challenge qualitative research as conventionally practiced.

How to Choose a Philosophy

After reading the previous list of philosophies—and perhaps the book as a whole—the question naturally arises: How should I, as a qualitative researcher, choose my philosophical position? This question is often raised more specifically in relation to concrete research methods: How should I choose which methods to employ? This latter question is often quite easy to answer because it depends on the research question with which one works. If one wants to study how people interact in everyday life, then ethnographic methods such as participant observation are needed. If one wants to study how people have experienced certain phenomena in their lives, then interview methods are often helpful. If one wants to study what happens in a social practice when something new occurs, then some form of action research might be needed. In relation to philosophy, however, the question is more difficult because one's basic philosophical orientation will often affect which research question one becomes interested in to begin with. Thus, there seems to be a certain a priori character to philosophy that does not apply to specific research methods.

Perhaps something is wrong with the question itself? For perhaps one does not "choose" a philosophy as such? I will not deny that it may happen, but I believe that more often one discovers that a certain philosophical outlook is already built into one's life and one's way of viewing things. Human beings can certainly make philosophy "a way of life," as Hadot (1995) argued, but all human lives are always already organized along philosophical ideas that are built into history, society, culture, and subculture—that is, ontological ideas about what exists, epistemological ideas about how one may come to know anything, and ethical ideas about how one ought to live. If this is true, then philosophizing is not primarily about inventing everything from scratch (although some philosophical positions, such as that of Descartes, have thought so) but, rather, about critically reflecting on and engaging with the ideas that have been formative in one's life. Most people do have positivist ideas about science as a practice that strives for truth through systematic observation. They also have realist ideas about a world beyond appearances; phenomenological ideas about openness to whatever happens in one's life; hermeneutic ideas about interpretation and the importance of getting clear about one's prejudices; pragmatist ideas about the significance of acting in practice; structuralist and post-structuralist ideas about meaning holism and the power of discourses; and also ideas about respect for those worldviews that have been sidelined by the dominant strands of Western thought. If so, one may be able to appreciate that all major philosophical positions ought to have a foot in the discussion, so to say, and none should be exorcised simply because of how they are conventionally represented (as I have argued, dismissing something because it is "positivist" is very common among qualitative researchers, and yet it is a quite prejudicial practice among many human scientists who rarely have in-depth knowledge about the positivist traditions).

Thus, this means that the practice of philosophy runs through the whole process of conducting qualitative research. Philosophy is not just a set of axioms or theses from which research methodologies and techniques may follow but, rather, a continuing practice of questioning, where certain traditions, such as those highlighted in this book, have developed specific genres of argument and reflection, but where they very often can be combined or at least borrow from each other in fruitful ways.

Making Philosophy Practical in Qualitative Research

The next question then becomes: How to do this concretely in a qualitative research project? Well, if it is fruitful to approach philosophy as a way of life—and thus also as a way of doing research because research is arguably a subset of human living (Brinkmann, 2012)—then philosophy cannot be confined to a specific phase in one's research process, or indeed to a specific section of one's research reports, when one writes an article or book about the research conducted. It may certainly be useful for the reader to read a section on post-structuralism, phenomenology, or whatever is one's favored philosophical paradigm, but this should not hide the fact that there are also philosophical ideas and arguments implicit in the design, methods, results, and discussion sections—and quite trivially in one's reflections on research ethics. Philosophy is not just a set of founding principles but also a continuing practice of questioning and critical self-reflection throughout any process of inquiry. How to do this can best be appreciated by looking at real, practical examples, and thus I urge the reader to revisit the chapters of the book and consult the exemplars of qualitative research that have been used as illustrations of the different philosophical ideas.

However, with the ambition of being more concrete, I end by briefly referring to four standard phases of a qualitative research project and for each phase discuss specific philosophical issues (the phases are discussed thoroughly with respect to interviewing specifically in Brinkmann, 2013b). The four phases are preparation, data collection, analysis, and reporting.

Preparation

The first phase crucially involves the formulation of a research question. Before one knows *what* to study, it is difficult to decide on *how*. All research questions are infused with philosophical presuppositions and implications. As discussed in this book, no words are philosophically innocent, but all are laden with ontological, epistemological, and ethical ideas. In a preparation phase, the qualitative researcher should therefore not only think about a good and relevant research question, or about how many interviewees to invite and which questions to ask, but also think through how implicit philosophical assumptions might affect the research process. Does the researcher work with a realist or anti-realist approach? What are the implications? Should the overall strategy be one of making the obvious obvious, making the hidden obvious, making the hidden dubious, or making the obvious dubious? Sometimes, as in phenomenology, a whole philosophical "package" is available and can be used, but also in these cases—and perhaps especially in these cases—a thorough philosophical reflection is needed so that one does not blindly work with a set of ideas that one really cannot defend. In the preparation phase, it is also relevant to think about alignment. An eclecticism that borrows from different philosophical paradigms can be useful if one can show that it is viable but unreflective eclecticism—for example, one that combines a constructionist epistemology with a phenomenological methodology without any argument is problematic.

Data Collection

It should be obvious now that even framing something as "data collection" is already to have taken a philosophical position. For non-realist positions, there is no such thing as data collection, but perhaps data creation, enactment, or translation. So, in the phase of gathering or constructing a corpus of empirical materials (through interviewing, participant observation, document collection, etc.), there are many philosophical questions. Each of the philosophical traditions discussed in this book has its take on what is meant by "data collection." In the British traditions of positivism and realism, the term will be relatively unproblematic, whereas hermeneutic and post-structural scholars will approach it with suspicion. In the recent "post-qualitative" movement, some have given up entirely on the distinction between "data" (as something arising from empirical materials) and "theory" (as something coming from books and articles that one has read) because the two are considered as being on the same ontological plane, so to speak (St. Pierre, 2011).

Analysis

The analysis phase, in which one sits and considers one's qualitative research materials, is perhaps when philosophy is most directly relevant. Positivism, phenomenology, and structuralism, for example, will offer quite concrete ways of addressing one's materials—what to look for and how to code and categorize—whereas realism, hermeneutics, and the post-structuralisms may enable the researcher to do a kind of "theoretical" reading in a more general sense. The different philosophical paradigms here differ concerning what they view as constituents of the world being studied, and again the distinction between realism and anti-realism becomes relevant (are one's categories, for example, reflections of something preexisting in the materials, or does the researcher create an order through the analysis that was not there before?). Also, the whole discussion about representation is relevant here: How can something—for example, the interpretations that one formulates about what participants in a research project have said—ever be said to represent or stand for what people meant? As we have seen, the different philosophical paradigms provide different answers to this question.

Reporting

Writing about one's qualitative research—or communicating about it through other media such as performance—is laden with the same philosophical issues as the other phases of research. The ethical problems, however, become particularly pertinent here because in qualitative research one often takes private stories and accounts and then places them in the public arena when reporting. Thus, participants can read about themselves, and others may recognize them. This requires in an obvious sense an "ethics of reporting," but a research report may affect the ways that important matters are understood and discussed in a more subtle sense, and it can therefore have an impact on the field under scrutiny. In the beginning of the book, I presented Aristotle's distinction between different kinds of knowledge (*episteme, techne,* and *phronesis*), and it is relevant to consider how one wants to understand the knowledge produced and communicated through a qualitative research project. Is it knowledge that is about a given reality that can be left unaffected by the process of knowing (*theoria*)? Is it knowledge that enables one to effectuate a desired change (*techne*)? Or is it a kind of ethico-political knowledge that enables a community to act with more wisdom in relation to significant questions (*phronesis*)? Or perhaps even a combination? How one decides to write up one's research should be partly informed by what one wants to achieve and how one considers the nature of the knowledge produced. This book is not about writing up qualitative research, but there are other sources that address exactly this topic (Brinkmann, 2013b; Wolcott, 2009).

REFERENCES

Adorno, T. W., Frenkel-Brunswik, E., Levinson, D. J., & Sanford, R. N. (1950). *The authoritarian personality*. New York, NY: Norton.
Aghasaleh, R., & St. Pierre, E. A. (2014). *A reader's guide to post-qualitative inquiry proposals*. Retrieved from https://goo.gl/3OC5b2
Ahenakew, C. (2016). Grafting indigenous ways of knowing onto non-indigenous ways of being: The (underestimated) challenges of a decolonial imagination. *International Review of Qualitative Research, 9*, 323–340.
Alvesson, M., & Kärreman, D. (2011). *Qualitative research and theory development: Mystery as method*. London, England: Sage.
Annas, J. (2001). Ethics and morality. In L. C. Becker & C. B. Becker (Eds.), *Encyclopedia of ethics* (2nd ed.). London, England: Routledge.
Aristotle. (1976). *Nichomachean ethics*. London, England: Penguin.
Arribas-Ayllon, M., & Walkerdine, V. (2008). Foucauldian discourse analysis. In C. Willig & W. Stainton-Rogers (Eds.), *The Sage handbook of qualitative research in psychology* (pp. 91–108). London, England: Sage.
Ayer, A. J. (1990). *Language, truth and logic*. London, England: Penguin. (Original work published 1936)
Barad, K. (2007). *Meeting the universe halfway: Quantum physics and the entanglement of matter and meaning*. Durham, NC: Duke University Press.
Beck, T. J. (2013). A phenomenological analysis of anxiety as experienced in social situations. *Journal of Phenomenological Psychology, 44*, 179–219.

Becker, H. (1967). Whose side are we on? *Social Problems, 14*, 239–247.
Bellah, R. N., Madsen, R., Sullivan, W. M., Swidler, A., & Tipton, S. M. (1985). *Habits of the heart: Individualism and commitment in American life*. Berkeley, CA: University of California Press.
Bennett, M. R., & Hacker, P. M. S. (2003). *Philosophical foundations of neuroscience*. Oxford, England: Blackwell.
Bhaskar, R. (2008). *A realist theory of science*. London, England: Routledge. (Original work published 1975)
Bishop, R. (2005). Freeing ourselves from neocolonial domination in research: A Kaupapa Maori approach to creating knowledge. In N. K. Denzin & Y. S. Lincoln (Eds.), *The Sage handbook of qualitative research* (3rd ed., pp. 109–138). Thousand Oaks, CA: Sage.
Bjerre, H. J. (2015). *Analysér! [Analyze!]*. Copenhagen, Denmark: Mindspace.
Blumer, H. (1969). *Symbolic interactionism: Perspective and method*. Englewood Cliffs, NJ: Prentice-Hall.
Boltanski, L., & Chiapello, E. (2005). *The new spirit of capitalism*. London, England: Verso.
Bourdeau, M., Braunstein, J.-F., & Petit, A. (2003). *Auguste Comte Aujourd'hui*. Paris, France: Kimé.
Bourdieu, P. (1977). *Outline of a theory of practice*. Cambridge, England: Cambridge University Press.
Brinkmann, S. (2004). Psychology as a moral science: Aspects of John Dewey's psychology. *History of the Human Sciences, 17*, 1–28.
Brinkmann, S. (2005a). Human kinds and looping effects in psychology: Foucauldian and hermeneutic perspectives. *Theory & Psychology, 15*, 769–791.
Brinkmann, S. (2005b). Psychology's facts and values: A perennial entanglement. *Philosophical Psychology, 18*, 749–765.
Brinkmann, S. (2007). Practical reason and positioning. *Journal of Moral Education, 36*, 415–432.
Brinkmann, S. (2008). Identity as self-interpretation. *Theory & Psychology, 18*, 404–422.
Brinkmann, S. (2010). Guilt in a fluid culture: A view from positioning theory. *Culture & Psychology, 16*, 253–266.
Brinkmann, S. (2011). *Psychology as a moral science: Perspectives on normativity*. New York, NY: Springer.
Brinkmann, S. (2012). *Qualitative inquiry in everyday life: Working with everyday life materials*. London, England: Sage.
Brinkmann, S. (2013a). *John Dewey: Science for a changing world*. New Brunswick, NJ: Transaction Publishers.
Brinkmann, S. (2013b). *Qualitative interviewing*. New York, NY: Oxford University Press.

Brinkmann, S. (2014). Psychiatric diagnoses as semiotic mediators: The case of ADHD. *Nordic Psychology, 66,* 121–134.

Brinkmann, S. (2017). Humanism after posthumanism: Or qualitative psychology after the "posts." *Qualitative Research in Psychology, 14*(2), 109–130.

Brinkmann, S., Jacobsen, M. H., & Kristiansen, S. (2014). Historical overview of qualitative research in the social sciences. In P. Leavy (Ed.), *The Oxford handbook of qualitative research* (pp. 17–42). Oxford, England: Oxford University Press.

Brinkmann, S., & Kvale, S. (2005). Confronting the ethics of qualitative research. *Journal of Constructivist Psychology, 18,* 157–181.

Brinkmann, S., & Kvale, S. (2015). *InterViews: Learning the craft of qualitative research interviewing* (3rd ed.). Thousand Oaks, CA: Sage.

Brinkmann, S., & Tanggaard, L. (2010). Toward an epistemology of the hand. *Studies in Philosophy and Education, 29,* 243–257.

Broden, T. F. (2010). Ferdinand de Saussure and linguistic structuralism. In D. Ingram (Ed.), *Critical theory to structuralism: Philosophy, politics, and the human sciences* (pp. 221–244). Durham, NC: Acumen.

Bryant, A. (2003). A constructive/ist response to Glaser. *Forum: Qualitative Social Research, 4*(1), Art. 15. Retrieved from http://nbn-resolving.de/urn:nbn:de:0114-fqs0301155

Cahan, E. D., & White, S. H. (1992). Proposals for a second psychology. *American Psychologist, 47,* 224–235.

Coles, R. (1992). *Self/power/other: Political theory and dialogical ethics.* Ithaca, NY: Cornell University Press.

Comte, A. (1988). *Introduction to positive philosophy.* Indianapolis: Hackett. (Original work published 1830)

Cornish, F., & Gillespie, A. (2009). A pragmatist approach to the problem of knowledge in health psychology. *Journal of Health Psychology, 14,* 800–809.

Coupland, D. (2009). *Generation A.* New York, NY: Scribner.

Critchley, S. (2007). *Infinitely demanding: Ethics of commitment, politics of resistance.* London, England: Verso.

Crowell, S. (2009). Husserlian phenomenology. In H. Dreyfus & M. Wrathall (Eds.), *A companion to phenomenology and existentialism* (pp. 9–30). Oxford, England: Wiley-Blackwell.

Dale, P. A. (1989). *In pursuit of a scientific culture: Science, art, and society in the Victorian age.* Madison, WI: University of Wisconsin Press.

Davies, B., & Harré, R. (1999). Positioning and personhood. In R. Harré & L. van Langenhove (Eds.), *Positioning theory: Moral contexts of intentional action.* Oxford, England: Blackwell.

Davies, W. (2015). *The happiness industry: How the government and big business sold us well-being.* London, England: Verso.

170 : REFERENCES

Denzin, N. K., & Lincoln, Y. S. (2011a). Introduction. In N. K. Denzin & Y. S. Lincoln (Eds.), *The Sage handbook of qualitative research* (4th ed., pp. 1–19). Thousand Oaks, CA: Sage.

Denzin, N. K., & Lincoln, Y. S. (2011b). Strategies of inquiry. In N. K. Denzin & Y. S. Lincoln (Eds.), *The Sage handbook of qualitative research* (4th ed., pp. 243–250). Thousand Oaks, CA: Sage.

Derrida, J. (1970). *Of grammatology*. Baltimore, MD: Johns Hopkins University Press.

Derrida, J. (2001). *On cosmopolitanism and forgiveness*. London, England: Routledge.

Dewey, J. (1896). The reflex arc concept in psychology. *Psychological Review, 3*, 357–370.

Dewey, J. (1925). *Experience and nature*. Chicago, IL: Open Court.

Dewey, J. (1930). *Human nature and conduct: An introduction to social psychology*. New York, NY: The Modern Library. (Original work published 1922)

Dewey, J. (1946). *The public and its problems—An essay in political inquiry*. Chicago, IL: Gateway Books. (Original work published 1927)

Dewey, J. (1960). *The quest for certainty*. New York, NY: Capricorn. (Original work published 1929)

Dewey, J. (1991). *How we think*. Amherst, NY: Prometheus. (Original work published 1910)

Dewey, J., & Bentley, A. (1960). *Knowing and the known*. Boston, MA: Beacon Press. (Original work published 1949)

Dilthey, W. (1977). *Descriptive psychology and historical understanding*. The Hague, the Netherlands: Nijhoff. (Original work published 1894)

Dreyfus, H. (1991). *Being-in-the-world—A commentary on Heidegger's Being and Time, Division I*. Cambridge, MA: MIT Press.

Dreyfus, H., & Taylor, C. (2015). *Retrieving realism*. Cambridge, MA: Harvard University Press.

Edwards, D. (2004). Analyzing racial discourse: The discursive psychology of mind–world relationships. In H. van den Berg & M. Wetherell (Eds.), *Analyzing race talk: Multidisciplinary perspectives on the research interview*. Cambridge, England: Cambridge University Press.

Flyvbjerg, B. (2001). *Making social science matter—Why social inquiry fails and how it can succeed again*. Cambridge, England: Cambridge University Press.

Fog, J. (2004). *Med samtalen som udgangspunkt* (2nd ed.). Copenhagen, Denmark: Akademisk Forlag.

Foucault, M. (1977). *Discipline and punish: The birth of the prison*. New York, NY: Pantheon.

Foucault, M. (1980). *The history of sexuality, Volume 1: An introduction*. New York, NY: Random House.

Foucault, M. (1984). On the genealogy of ethics: An overview of work in progress. In P. Rabinow (Ed.), *The Foucault reader*. London, England: Penguin.
Foucault, M. (1988a). Technologies of the self. In L. Martin, H. Gutman & P. Hutton (Eds.), *Technologies of the Self*. London, England: Tavistock.
Foucault, M. (1988b). Truth, power, self. In L. Martin, H. Gutman & P. Hutton (Eds.), *Technologies of the self*. London, England: Tavistock.
Foucault, M. (1990). *The history of sexuality, Vol. 3: The care of the self*. London, England: Penguin.
Foucault, M. (1994). The subject and power. In J. D. Faubion (Ed.), *Power: Essential works of Michel Foucault, Vol. 3*. London, England: Penguin.
Foucault, M. (2001). *The order of things: An archeology of the human sciences*. London, England: Taylor & Francis. (Original work published 1966)
Gadamer, H. G. (2000). *Truth and method*. (Second revised edition published 2000). New York, NY: Continuum. (Original work published 1960)
Garfinkel, H. (1984). *Studies in ethnomethodology*. Cambridge, England: Polity. (Original work published 1967)
Garrison, J. (1999). John Dewey's theory of practical reasoning. *Educational Philosophy and Theory, 31*, 291–312.
Gazzaniga, M., & Heatherton, T. (2003). *Psychological science: Mind, brain and behavior*. New York, NY: Norton.
Gergen, M. (2008). Qualitative methods in feminist psychology. In C. Willig & W. Stainton-Rogers (Eds.), *The Sage handbook of qualitative research in psychology* (pp. 280–295). London, England: Sage.
Gibbs, G. (2007). *Analyzing qualitative data*. London, England: Sage.
Giddens, A. (1976). *New rules of sociological method*. London, England: Hutchinson.
Gier, N. F. (1981). *Wittgenstein and phenomenology: A comparative study of the later Wittgenstein, Husserl, Heidegger and Merleau-Ponty*. Albany, NY: State University of New York Press.
Gilligan, C. (1982). *In a different voice*. Cambridge, MA: Harvard University Press.
Gilligan, C., Spencer, R., Weinberg, M. K., & Bertsch, T. (2003). On the listening guide: A voice-centered relational model. In P. M. Camic, J. E. Rhodes, & L. Yardley (Eds.), *Qualitative research in psychology: Expanding perspectives in methodology and design*. Washington, DC: American Psychological Association.
Giorgi, A. (1975). An application of phenomenological method in psychology. In A. Giorgi, C. Fischer, & E. Murray (Eds.), *Duquesne*

studies in phenomenological psychology II. Pittsburgh, PA: Duquesne University Press.

Glaser, B. G., & Strauss, A. (1967). *The discovery of grounded theory: Strategies for qualitative research*. New York, NY: Aldine.

Goffman, E. (1959). *The presentation of self in everyday life*. New York, NY: Overlook Press.

Hacking, I. (1995). The looping effect of human kinds. In D. Sperber, D. Premack, & A. J. Premack (Eds.), *Causal cognition: A multidisciplinary debate*. Oxford, England: Clarendon.

Haddad, S. (2010). Jacques Derrida. In A. Schrift (Ed.), *Poststructuralism and critical theory's second generation* (pp. 111–132). Durham, NC: Acumen.

Hadot, P. (1995). *Philosophy as a way of life: Spiritual exercises from Socrates to Foucault*. Oxford, England: Blackwell.

Hammersley, M. (2008). *Questioning qualitative inquiry: Critical essays*. London, England: Sage.

Hanson, N. R. (1969). *Perception and discovery: An introduction to scientific inquiry*. San Francisco, CA: Freeman, Cooper & Co.

Haraway, D. (1991). *Simians, cyborgs and women: The reinvention of nature*. New York, NY: Routledge.

Harré, R. (1997). Forward to Aristotle: The case for a hybrid ontology. *Journal for the Theory of Social Behaviour, 27*, 173–191.

Harré, R. (2002). Social reality and the myth of social structure. *European Journal of Social Theory, 5*, 111–123.

Harré, R., & Madden, E. H. (1975). *Causal powers: Theory of natural necessity*. Oxford, England: Blackwell.

Harré, R., & Moghaddam, F. M. (2003). Introduction: The self and others in traditional psychology and in positioning theory. In R. Harré & F. M. Moghaddam (Eds.), *The self and others: Positioning individuals and groups in personal, political, and cultural contexts* (pp. 1–11). London, England: Praeger.

Harré, R., & van Langenhove, L. (1999). The dynamics of social episodes. In R. Harré & L. van Langenhove (Eds.), *Positioning theory* (pp. 1–13). Oxford, England: Blackwell.

Hegel, G. W. F. (1977). *Phenomenology of spirit*. Oxford, England: Oxford University Press. (Original work published 1807)

Heidegger, M. (1962). *Being and time*. New York, NY: HarperCollins. (Original work published 1927)

Herzog, M. (1995). William James and the development of phenomenological psychology in Europe. *History of the Human Sciences, 8*, 29–46.

hooks, b. (2000). *Feminist theory: From margin to center*. Cambridge, MA: South End Press.

Hume, D. (1978). *A treatise of human nature: Being an attempt to introduce the experimental method of reasoning into moral subjects.* Oxford, England: Clarendon. (Original work published 1739)

Husserl, E. (1954). *Die Krisis der europäischen Wissenschaften und die tranzendentale Phänomenologie.* The Hague, the Netherlands: Martinus Nijhoff.

Ingram, D. (2010). Introduction. In D. Ingram (Ed.), *Critical theory to structuralism: Philosophy, politics, and the human sciences* (pp. 1–18). Durham, NC: Acumen.

James, W. (1981). *Pragmatism.* Indianapolis, IN: Hackett. (Original work publishes 1907)

Jefferson, G. (1985). An exercise in the transcription and analysis of laughter. In T. Van Dijk (Ed.), *Handbook of discourse analysis* (Vol. 3). London, England: Academic Press.

Joas, H. (1996). *The creativity of action.* Cambridge, England: Polity.

Kessen, W., & Cahan, E. D. (1986). A century of psychology: From subject to object to agent. *American Scientist, 74,* 640–649.

Kestenbaum, V. (1977). *The phenomenological sense of John Dewey: Habit and meaning.* Atlantic Highlands, NJ: Humanities Press.

Kohlberg, L. (1981). *Essays on moral development Volume 1—The philosophy of moral development.* San Francisco, CA: Harper & Row.

Köhler, W. (1959). *The place of value in a world of facts.* New York, NY: Meridian. (Original work published 1938)

Kuhn, T. S. (1970). *The structure of scientific revolutions* (2nd enlarged ed.). Chicago, IL: University of Chicago Press.

Kvale, S. (2006). Dominance through interviews and dialogues. *Qualitative Inquiry, 12,* 480–500.

Kvale, S. (2008). Qualitative inquiry between scientific evidentialism, ethical subjectivism and the free market. *International Review of Qualitative Research, 1,* 5–18.

Kvale, S. (1992). Postmodern psychology: A contradiction in terms? In S. Kvale (Ed.), *Psychology and postmodernism* (pp. 31–57). Thousand Oaks, CA: Sage.

Langdridge, D. (2007). *Phenomenological psychology: Theory, research and method.* Harlow, England: Pearson.

Lather, P. (2006). Paradigm proliferation as a good thing to think with: Teaching research in education as a wild profusion. *International Journal of Qualitative Studies in Education, 19,* 35–57.

Lather, P. (2016). Top ten+ list: (Re)thinking ontology in (post)qualitative research. *Cultural Studies—Critical Methodologies, 16,* 125–131.

Lather, P., & St. Pierre, E. A. (2013). Post-qualitative research. *International Journal of Qualitative Studies in Education, 26,* 629–633.

Latour, B. (1993). *We have never been modern.* Cambridge, MA: Harvard University Press.
Latour, B. (2005). *Reassembling the social.* Oxford, England: Oxford University Press.
Leplin, J. (2007). Enlisting Popper in the case for scientific realism. *Studies in History of Sciences and Philosophy, 11,* 71–97.
Levinas, E. (1969). *Totality and infinity: An essay on exteriority.* Pittsburgh, PA: Duquesne University Press.
Lincoln, Y. S., Lynham, S. A., & Guba, E. G. (2011). Paradigmatic controversies, contradictions, and emerging confluences, revisited. In N. K. Denzin & Y. S. Lincoln (Eds.), *The Sage handbook of qualitative research* (4th ed., pp. 97–128). Thousand Oaks, CA: Sage.
Lindner, R. (1996). *The reportage of urban culture: Robert Park and the Chicago School.* Cambridge, England: Cambridge University Press.
Lyotard, J.-F. (1984). *The postmodern condition: A report on knowledge.* Manchester, England: Manchester University Press.
MacLure, M. (2013). Classification or wonder? Coding as an analytic practice in qualitative research. In R. Coleman & J. Ringrose (Eds.), *Deleuze and research methodologies* (pp. 164–183). Edinburgh, England: Edinburgh University Press.
MacLure, M. (2016). *Qualitative methodology and the new materialisms: Do we need a new conceptual vocabulary?* Keynote address May 19, 2016, at the 12th International Congress of Qualitative Inquiry at the the University of Illinois at Urbana–Campaign.
Malcolm, M. (1988). *Nothing is hidden.* Oxford, England: Wiley-Blackwell.
Marx, K. (1894). *Capital: Critique of political economy, Vol. III.* Retrieved from https://www.marxists.org/archive/marx/works/1894-c3/ch48.htm
Marx, K. (1888). *Theses on Feuerbach.* Retrieved from https://www.marxists.org/archive/marx/works/1845/theses/theses.htm
Maxwell, J. (2012). *A realist approach to qualitative research.* Thousand Oaks, CA: Sage.
May, T. (2009). *Death.* Durham, NC: Acumen.
McDowell, J. (1998). *Mind, value, and reality.* Cambridge, MA: Harvard University Press.
Mead, G. H. (1962). *Mind, self, and society: From the standpoint of a social behaviorist.* Chicago, IL: University of Chicago Press. (Original work published 1934)
Menand, L. (2002). *The metaphysical club.* London, England: Flamingo.
Merleau-Ponty, M. (2002). *Phenomenology of perception.* London, England: Routledge. (Original work published 1945)
Michell, J. (2003). The quantitative imperative: Positivism, naïve realism and the place of qualitative methods in psychology. *Theory & Psychology, 13,* 5–31.

Mill, J. S. (1987). *The logic of the moral sciences*. London, England: Duckworth. (Original work published 1843)

Miller, R. B. (2004). *Facing human suffering: Psychology and psychotherapy as moral engagement*. Washington, DC: American Psychological Association.

Mills, C. W. (2000). *The sociological imagination*. Oxford, England: Oxford University Press. (Original work published 1959)

Morgan, D. L. (2014). Pragmatism as a paradigm for social research. *Qualitative Inquiry, 20*, 1045–1053.

Nagel, T. (1986). *The view from nowhere*. Oxford, England: Oxford University Press.

Neurath, O. (2003). The scientific world conception. In G. Delanty & P. Strydom (Eds.), *Philosophies of social science: The classic and contemporary readings* (pp. 31–34). Maidenhead, England: Open University Press. (Original work published 1929)

Niiniluoto, I. (1996). Queries about internal realism. In R. S. Cohen, R. Hilpinen, & R.-Z. Qiu (Eds.), *Realism and anti-realism in the philosophy of science* (pp. 45–54). Dordrecht, the Netherlands: Springer.

Noblit, G. W., & Hare, R. D. (1988). *Meta-ethnography: Synthesizing qualitative studies*. Newbury Park, CA: Sage.

Norris, C. (1987). *Derrida*. London, England: Fontana.

Nussbaum, M. C. (1986). *The fragility of goodness: Luck and ethics in Greek tragedy and philosophy*. Cambridge, England: Cambridge University Press.

Okrent, M. (1988). *Heidegger's pragmatism: Understanding, being, and the critique of metaphysics*. Ithaca, NY: Cornell University Press.

Olesen, V. (2011). Feminist qualitative research in the millennium's first decade. In N. K. Denzin & Y. S. Lincoln (Eds.), *The Sage handbook of qualitative research* (4th ed., pp. 129–146). Thousand Oaks, CA: Sage.

Parker, I. (2005). *Qualitative psychology: Introducing radical research*. Buckingham, England: Open University Press.

Pelias, R. (2004). *A methodology of the heart: Evoking academic and daily life*. Walnut Creek, CA: AltaMira.

Plato. (1987). *The republic*. London, England: Penguin.

Polkinghorne, D. (2000). Psychological inquiry and the pragmatic and hermeneutic traditions. *Theory & Psychology, 10*, 453–479.

Polkinghorne, D. (2004). *Practice and the human sciences*. Albany, NY: State University of New York Press.

Popper, K. R. (1959). *The logic of scientific discovery*. London, England: Routledge.

Potter, J., & Hepburn, A. (2005). Qualitative interviews in psychology: Problems and possibilities. *Qualitative Research in Psychology, 2*, 281–307.

Potter, J., & Hepburn, A. (2008). Discursive constructionism. In J. A. Holstein & J. F. Gubrium (Eds.), *Handbook of constructionist research* (pp. 275–293). New York, NY: Guilford.
Potter, J., & Wetherell, M. (1987). *Discourse and social psychology*. London, England: Sage.
Putnam, H. (1994). Pragmatism and moral objectivity. In J. Conant (Ed.), *Words and life*. Cambridge, MA: Harvard University Press.
Putnam, H. (2002). *The collapse of the fact/value dichotomy and other essays*. Cambridge, MA: Harvard University Press.
Quine, W. V. (1951). Two dogmas of empiricism. *Philosophical Review*, 60, 20–43.
Qvortrup, L. (2003). *The hypercomplex society*. New York, NY: Lang.
Radnitzky, G. (1970). *Contemporary schools of metascience*. Göteborg, Sweden: Akademiforlaget.
Ray, L. (2016). "Beading becomes a part of your life": Transforming the academy through the use of beading as a method of inquiry. *International Review of Qualitative Research*, 9, 363–378.
Richardson, F. C., Fowers, B. J., & Guignon, C. B. (1999). *Re-envisioning psychology: Moral dimensions of theory and practice*. San Francisco, CA: Jossey-Bass.
Richardson, L., & St. Pierre, E. A. (2005). Writing: A method of inquiry. In N. K. Denzin & Y. S. Lincoln (Eds.), *Handbook of qualitative research* (3rd ed., pp. 959–978). Thousand Oaks, CA: Sage.
Ricoeur, P. (1991). Life in quest of narrative. In D. Wood (Ed.), *On Paul Ricoeur: Narrative and interpretation* (pp. 20–33). London, England: Routledge.
Robinson, D. N. (2008). *Consciousness and mental life*. New York, NY: Columbia University Press.
Rorty, R. (1980). *Philosophy and the mirror of nature*. Princeton, NJ: Princeton University Press.
Rorty, R. (1982). *Consequences of pragmatism*. Brighton, England: Harvester.
Rorty, R. (1989). *Contingency, irony, and solidarity*. Cambridge, England: Cambridge University Press.
Samelson, F. (1974). History, origin myth and ideology: "Discovery" of social psychology. *Journal for the Theory of Social Behaviour*, 4, 217–231.
Sartre, J.-P. (1966). *Being and nothingness*. New York, NY: Washington Square Press. (Original work published 1943)
Schrift, A. (2010). Introduction. In A. Schrift (Ed.), *Poststructuralism and critical theory's second generation* (pp. 1–17). Durham, NC: Acumen.
Schwandt, T. (2000). Three epistemological stances for qualitative inquiry: Interpretivism, hermeneutics, and social constructionism. In

N. K. Denzin & Y. S. Lincoln (Eds.), *Handbook of qualitative research* (pp. 189–213). London, England: Sage.

Schwandt, T. (2001). *Dictionary of qualitative inquiry* (2nd ed.). Thousand Oaks, CA: Sage.

Sellars, W. (1997). *Empiricism and the philosophy of mind*. Cambridge, MA: Harvard University Press. (Original work published 1956)

Sharma, A. (2013). Hinduism. In C. Meister & P. Copan (Eds.), *The Routledge companion to the philosophy of religion* (2nd ed., pp. 7–17). London, England: Routledge.

Singer, B. C. J. (2010). Claude Lévi-Strauss. In D. Ingram (Ed.), *Critical theory to structuralism: Philosophy, politics, and the human sciences* (pp. 245–261). Durham, NC: Acumen.

Smith, J. A., Flowers, P., & Larkin, M. (2009). *Interpretative phenomenological analysis*. London, England: Sage.

Smith, R. (1997). *The Norton history of the human sciences*. New York, NY: Norton.

Spencer, R., Pryce, J. M., & Walsh, J. (2014). Philosophical approaches to qualitative research. In P. Leavy (Ed.), *The Oxford handbook of qualitative research* (pp. 81–98). Oxford, England: Oxford University Press.

St. Pierre, E. A. (2008). Decentering voice in qualitative inquiry. *International Review of Qualitative Research*, *1*, 319–336.

St. Pierre, E. A. (2011). Post qualitative inquiry. In N. K. Denzin & Y. S. Lincoln (Eds.), *The Sage handbook of qualitative research* (4th ed., pp. 611–625). Thousand Oaks, CA: Sage.

St. Pierre, E. A., Jackson, A. Y., & Mazzei, L. (2016). New empiricisms and new materialisms: Conditions for new inquiry. *Cultural Studies—Critical Methodologies*, *16*, 99–110.

Tachine, A. R., Yellow Bird, E., & Cabrera, N. L. (2016). Sharing circles: An indigenous methodological approach for researching with groups of indigenous peoples. *International Review of Qualitative Research*, *9*, 277–295.

Taylor, C. (1985a). Self-interpreting animals. In *Human agency and language: Philosophical papers 1* (pp. 45–76). Cambridge, England: Cambridge University Press.

Taylor, C. (1985b). What is human agency? In *Human agency and language: Philosophical papers 1* (pp. 15–44). Cambridge, England: Cambridge University Press.

Taylor, C. (1995). Overcoming epistemology. In *Philosophical arguments* (pp. 1–19). Cambridge, MA: Harvard University Press.

Thorne, S. E. (2014). Applied interpretive approaches. In P. Leavy (Ed.), *The Oxford handbook of qualitative research* (pp. 99–115). Oxford, England: Oxford University Press.

Toulmin, S. (1990). *Cosmopolis: The hidden agenda of modernity.* Chicago, IL: University of Chicago Press.

Toulmin, S. (2001). *Return to reason.* Cambridge, MA: Harvard University Press.

van Kaam, A. (1959). Phenomenal analysis: Exemplified by a study of the experience of "really feeling understood." *Journal of Individual Psychology, 15,* 66–72.

Venable, J., Sato, B. A., Del Duca, J., & Sage, F. (2016). Decolonizing our own stories: A project of the student storytellers indigenizing the Academy (SSITA) group. *International Review of Qualitative Research, 9,* 341–362.

Wittgenstein, L. (1953). *Philosophical investigations.* Oxford, England: Basil Blackwell.

Wolcott, H. (2009). *Writing up qualitative research* (3rd ed.). Thousand Oaks, CA: Sage.

Wrathall, M. (2009). Existential phenomenology. In H. Dreyfus & M. Wrathall (Eds.), *A companion to phenomenology and existentialism* (pp. 31–47). Oxford, England: Wiley-Blackwell.

Young, I. M. (1980). Throwing like a girl: A phenomenology of feminine body comportment, motility and spatiality. *Human Studies, 3,* 137–156.

INDEX

Note: Page numbers followed by italicized letters indicate text found in boxes (*b*) and tables (*t*).

A

abduction, 103–104, 110, 113
acts, in positioning theory, 135–136
actual domain, 58–59
Addams, Jane, 105
Adorno, T. W., 62–63
aesthetics, 7, 17
affectedness, 70
Aghasaleh, R., 149
Ahenakew, C., 144–145
Althusser, Louis, 116, 118–119
American philosophies of qualitative research. *See* pragmatism
analysis. *See* data collection and analysis
Anselm of Canterbury, 30
anthropocentric cosmology, 32, 35, 43
anti-realism
  as positivist stance, 48, 56, 61–63
  vs. realism, 8–9, 10, 31, 56, 64, 113, 155, 164

antiskepticism, 101
anxiety, study of, 84–86*b*
aporia, 3, 122
Aristotle
  Christianity synthesized with philosophy of, 29
  conception of nature, 35
  contemporary reprisals of, 31
  forms of knowledge, 13–15, 14*t*, 25–26
  as objectivist, 26–27, 28
  on political nature of humans and social sciences, 12–13, 27
  on problem of universals, 30
  rationality types, 27
  soul as matter of living body, 24–25
  teleological worldview, 26–27, 28–29
Arribas-Ayllon, M., 130, 131
art, philosophies represented in, 31–33, 50–51

articulation/telling, 70
authoritarian personalities, study of, 62–63*b*
Ayer, A. J., 53

B

Barad, Karen, 147, 148, 153*b*
Baudrillard, Jean, 129–130
Beauvoir, Simone de, 140
Beck, T. J., 84–86*b*
Becker, Howard, 12, 105
behaviorist science, critique of, 103
Being and Time (Heidegger), 70, 76, 92
Bellah, R. N., 15
Bennett, M. R., 37
Bentham, Jeremy, 126
Berkeley, George, 37–38
Bhaskar, Roy, 58
Bishop, R., 143–144
Bjerre, H. J., 17–18, 108
Blumer, Herbert, 87, 105–106
body, phenomenology of, 69, 70–71, 72, 83
body–soul/mind relationship, 24–25, 36, 39, 126
Bohr, Niels, 147, 148
bracketing, 80, 89
"breaching experiments," 109–110
Brinkmann, Svend, 15, 83–88, 123*b*
British philosophies of qualitative research, 102. *See also* positivism; realism
Broden, T. F., 118

C

Cahan, E. D., 38
Camus, Albert, 115
CAQDAS (computer-assisted qualitative data analysis software), 60, 63–64
Carmen (opera), 51
Cartesianism, 34–35, 93–94
causality
  Aristotle's view of, 27–28
  critical realist approaches to, 59, 61
  positivist approaches to, 39, 49, 56, 63, 157
  in psychologism, 68
  realist approaches to, 56, 61
  in scientific psychology, 52
cave allegory (Plato), 23, 44–45, 93
Cavell, Stanley, 4
central themes, *vs.* natural units, 83
change, *vs.* permanence, 22–23
Chicago School of Sociology, 104–105
Chomsky, Noam, 118
Christianity, 24, 29
coding, deconstructive reading of, 151
Cogito ergo sum, 36
cognitive science, 35, 45
colonizing narratives, 143, 144
Comte, Auguste, 48, 49–50, 52, 54
constructivism, 41
contemplation, 27
conversation analysis, 131
Cooley, Charles, 105
Cornish, F., 111–112*b*
correspondence theory of truth, 111*b*
cosmologies, 32, 35, 145, 154
Coupland, Douglas, 77
Course in General Linguistics (Saussure), 117
Critchley, Simon, 3–4, 121
critical rationalism, 56
critical realism. *See also* realism
  analytic strategies in, 17
  causality as viewed in, 59, 61
  data collection/analysis in, 61, 157–159, 164
  as dialectical theory, 58
  hidden made obvious in, 61
  key figures
    Bhaskar, 58
    Maxwell, 60–61
  ontology for, 58–59, 60, 157
  in qualitative research, 61
critical research, hidden made obvious by, 108
critical theory, 62*b*, 64
Critique of Dialectical Reason (Sartre), 74
Crowell, S., 66–67, 68

D

Darwinism, 98
Dasein, 70, 71
data collection and analysis
  in abduction process, 103
  CAQDAS, 60, 63–64
  critical realist approaches to, 61, 157–159
  discourse analysis, 130–134
  philosophical influences on, 163–164
  positivist approaches to, 60, 61
da Vinci, Leonardo, 32
death, philosophy as training for, 4
deconstruction
  defined, 18
  ethics of, 121–122
  Heidegger's influence on, 70
  of interviews, 123–124b
  obvious made dubious by, 123–124b, 137
  of ontology, 147, 150, 153b, 154
  as post-structuralist strategy, 119, 120–122, 136, 137, 160
  of representationalism, 150
deduction, *vs.* abduction, 103, 113
definitive concepts, 106
Deleuze, Gilles, 150, 151, 152, 153b
Democritus, 22
Denzin, Norman, 5–6, 105, 114
Derrida, Jacques, 18, 70, 120–122, 136
Descartes, Rene
  body–soul interaction, 36
  Cartesianism, 34–35, 36, 39, 93–94
  humans as "thinking things," 36
  ideas as representations of the world, 35, 94
  method of systematic doubt, 35–36
  as rationalist, 37
descriptive psychology, 76
Destruktion (Derrida), 120
Dewey, John
  abduction procedure, 103–104, 110
  on active nature of stimuli, 99
  dualisms opposed by, 96, 98–99, 102
  as founder of pragmatism, 92, 100
  ideas as viewed by, 94–95, 99–100
  matter as process, 148
  phenomenological ties, 92
  response to Marxism, 95–96
  Rorty's presentation of, 101
  on theory *vs.* practice, 93
dialogical method, 3
Dilthey, Wilhelm, 75–76, 88
disappointment, as beginning of philosophy, 3, 4
Discipline and Punish (Foucault), 125–126
discourse analysis, 130–134
discursive psychology, 132–133b, 134
double hermeneutics, 9
doubt, in abduction process, 103
doubt, systematic, 35–36
Dreyfus, H., 71, 72–73
Durkheim, Emile, 50–51
dynamic nominalism, 9

E

Edwards, D., 132–133b
efficient cause, 27–28, 35
eidos, 68
elenchus, 3
emotions
  hermeneutic views of, 77
  phenomenological views of, 67, 84–86b
  post-structuralist views of, 136
empirical domain, 58–59
empiricism, 57, 69, 131, 147
episteme, 13, 14, 126, 127–128. See also *theoria*
epistemological hermeneutics, 76
epistemology
  cosmological shift corresponding to, 32
  defined, 10
  in defining qualitative research, 7
  individualism and, 43
  in post-qualitative research, 150
  pragmatic observations about, 99
  as problem for qualitative research, 11
  realist approaches to, 10, 58, 60–61

epistemology (*cont.*)
  Taylor's argument for overcoming, 10–11
  as theoretical subdiscipline of philosophy, 7
  "epistemology of the eye," 44, 45
"equipment" (Heidegger), 71
Erasmus of Rotterdam, 34
Essais (Montaigne), 34
Essay Concerning Human Understanding (Locke), 37
essays, research findings presented in, 107
esse est percipi (to be is to be perceived), 37
ethico-onto-epistemology, 150
ethics
  of deconstruction, 121–122
  defined, 11
  morality related to, 42
  phronesis represented in, 14–15
  in post-qualitative research, 150
  as practical subdiscipline of philosophy, 7, 11–12
  as practice of freedom, 128
  in qualitative research practices, 7, 12, 151, 165
Ethics (Aristotle), 13, 27
ethnomethodologists, 108–109
eudaimonia/eudaemon, 26, 27
existential phenomenology. *See* phenomenology, existential
experience
  existential phenomenological views of, 69, 72
  phenomenological description of, 65–66, 67, 68, 69, 79, 152
  positivist views of, 49, 157
  post-qualitative views of, 152
  pragmatic views of, 93–94, 99, 113–114
Experience and Nature (Dewey), 92, 148

F

facts, as value-loaded, 12, 102
fallibilism, 101–102
falsification, *vs.* verfication, 56
feminist philosophies, 51, 140–141, 154
final cause, 27, 28
flat methodology, 149
flattened ontology, 149
Flyvbjerg, B., 15
forgiveness, deconstructive reading of, 121–122
formal cause, 27
Foucault, Michel
  creating self as work of art, 128–129
  genealogical method, 43
  historical contexts studied by, 125
  historical emergence of epistemes, 127–128
  modernity critiqued by, 125
  Nietzsche's influence on, 43
  qualitative research methods influenced by, 130–134, 136
  technologies for individual domination, 127–128
Frankfurt school of critical theory, 64
freedom, ethics as practice of, 128
Frege, Gottlob, 117
French philosophies of qualitative research. *See* post-structuralism; structuralism
fundamental ontology, 76

G

Gadamer, Hans-Georg, 76–80, 88, 159
Galileo, 29, 35
Garfinkel, Harold, 108–110
gender studies, 81–83
Generation A (Coupland), 77
Gergen, Mary, 141
German philosophies of qualitative research. *See* hermeneutics; phenomenology
Giddens, A., 9
Gier, N. F., 74
Gillespie, A., 111–112*b*
Gilligan, Carol, 141–142
Giorgi, Amedeo, 80, 83, 84*b*, 88
given, myth of, 152
Glaser, Barney, 60

global/local philosophies of
 qualitative research, 142–146,
 153b, 154, 160
God, philosophical views of, 3, 36, 38
Goffman, Erving, 106–108
Greek philosophy
 artistic representation of, 31–32
 key figures
  Aristotle, 24–29
  early philosophers, 22
  Plato. *See* Plato
 metaphysical speculation in, 22, 28
 revived in Renaissance
  philosophy, 31–32
 structure of reality as focus of, 10
Grotius, Hugo, 33
grounded theory, 60, 104
Guba, E. G., 156

H

habits, pragmatic views of, 98
Habits of the Heart (Bellah, et. al), 15
Hacker, P. M. S., 37
Hacking, Ian, 9, 59
Hadot, Pierre, 2, 64, 89, 161
Hanisch, Carol, 140
Hanson, N. R., 57, 58
Haraway, D., 148
Hare, R. D., 16, 48
Harré, R., 27, 134–135
Hegel, G. W. F., 41–43, 120
Heidegger, Martin
 Dasein, 70
 deconstructionists
  influenced by, 70
 hermeneutics influenced by, 75, 79,
  88, 159
 Kierkegaard's influence on, 43
 ontological hermeneutics, 76
 phenomenology influenced by,
  88, 159
 political views of, 70
 practical engagement as condition
  for experience, 69
 pragmatism influenced by, 92
 "present-at-hand"/"ready-to-hand,"
  71, 72, 120

 understanding as condition of
  being human, 76
Hepburn, A., 152–153
Heraclitus, 22, 23, 148
hermeneutical circle, 83–87
hermeneutic anthropology, 77
hermeneutics
 Aristotelian references in, 28
 canon of interpretations, 83–88
 data collection/analysis in, 164
 defined, 75
 epistemological *vs.* ontological, 76
 as German tradition, 17–18,
  66, 75, 91
 Heidegger's influence on, 75, 79
 as interpretation of text *vs.* life, 75
 key figures
  Dilthey, 75–76, 88
  Gadamer, 76–80, 159
  Schleiermacher, 75, 88
  Taylor, 76, 77, 78
 making the obvious obvious,
  17–18, 75, 82
 parts–whole iterative process
  in, 83–86
 vs. phenomenology, 88, 159
 as practical philosophy, 79–80
 in qualitative research, 78–79, 80,
  83–88, 161
hermeneutics, triple, 80
hermeneutics of suspicion, 18
hidden, making dubious
 as analytic strategy, 17, 18–19
 in pragmatism, 18–19, 108, 111–
  112b, 113, 159
hidden, making obvious
 as analytic strategy, 16, 17
 in critical research, 48, 108
 in realism, 56, 61, 62–63b, 64, 159
Holmes, Oliver Wendell, 92
hooks, bell, 141
Houellebecq, Michel, 51
"How to Make Our Ideas Clear"
 (Peirce), 97
How We Think (Dewey),
 103–104, 110
Hughes, Everett, 105

humanism
  dismantling of, 150–151, 160
  roots and relevance of, 32, 33–34
humans, philosophical paradigms for, 36, 77–78, 125–126, 139
Hume, David
  causality analyzed by, 39, 157
  as grandfather of positivism, 48–49
  is/ought dichotomy, 28, 39–40
  Kant's criticism of, 40, 41
  method-based approach, 38–39
  morality as unknowable, 39
  science of the mind, 39, 40
  self as illusion, 38
Hume's law, 39–40
Husserl, Edmund
  criticism of, 69
  intentionality as key concept for, 68
  lifeworld concept, 66, 71, 72
  phenomenological philosophy developed by, 51, 66–69, 79, 80, 84b, 88
  psychologism criticized by, 67–68
hypothesis, in abduction process, 103–104

I
idealism, 8, 30
"ideal participant," 86b
ideas
  Cartesian views of, 93–94
  cognitive science understanding of, 45
  Greek philosophical views of, 24, 30, 37–38, 44, 93
  Marxist views of, 95–96
  modern philosophical views of, 35, 36, 37, 38, 39, 43, 45
  pragmatic views of, 92–96, 99–100
In a Different Voice (Gilligan), 141–142
incongruence procedures, 109–110
indigenous philosophies, 142–146, 154
individualism, 43
induction, 55–56, 103, 113
instrumentalist theory of truth, 97–98, 101
intentionality, 68, 69, 76–77
interactionism, 105–106
interpellation, 119
interpretation, hermeneutic views of, 75–78, 86–88
intertextuality, 119
interviews
  deconstruction of, 123–124b, 151
  dialogic, 124b
  discursive psychology methodology for, 132–133b
  manifest vs. underlying questions, 62b
  qualitative psychology critique of, 152–154
is/ought dichotomy, 28, 39–40

J
Jackson, A. Y., 147–148
James, William, 92, 97–98, 100, 101

K
Kant, Immanuel, 40–42, 120
Kessen, W., 38
Kierkegaard, Søren, 43
Knowing and the Known (Dewey), 148
knowledge
  abduction process for investigating, 103–104
  critical realist approaches to, 59
  Descartes' views of, 35–36
  epistemology as theory of, 10
  ethics of reporting significant for, 165
  feminist views of, 141, 154
  forms of, 14–15, 14t
  hermeneutical views of, 77, 78, 79–80, 87
  indigenous paradigms for, 145–146
  Kant's view of, 41
  local, 32–33
  philosophical pursuit of, 2
  positivist approaches to, 49, 157
  postmodern views of, 32–33, 129
  post-qualitative approaches to, 146

pragmatic views of, 10, 91, 100, 106, 111–112*b*, 114
scientific. *See* scientific knowledge
stages of, 49
virtues of, 14, 14*t*, 25–26
Kuhn, T. S., 57–58
Kvale, Steinar, 15, 83–88, 123*b*

L

Lacan, Jacques, 119
language
　as embodied, practical activity, 74
　postmodern views of, 129
　pragmatic views of, 101
　structuralist views of, 117–118, 119, 121, 137
Language, Truth and Logic (Ayer), 53
langue, 117
Lather, Patti, 147, 153*b*, 157
Latour, Bruno, 148, 150
Levinas, E., 21
Lévi-Strauss, Claude, 115–116, 118, 120, 136
lifeworld, 66, 71, 72, 74, 75
Lincoln, Yvonna, 5–6, 114, 156
linguistics, as science, 117
linguistics, structural, 118
local knowledges, 32–33
Locke, John, 37, 38
Logic of the Moral Sciences, The (Mill), 51
looping effect, 9
Lynham, S. A., 156
Lyotard, Jean-François, 129

M

MacLure, Maggie, 142, 151
Madame Bovary (Flaubert), 51
Madsen, R., 15
Malcolm, M., 74–75
Maori, 143–144
Martineau, Harriet, 51
Marx, Karl. *See also* Marxism
　activist approach of, 42
　on critical nature of science, 17, 58
　ideas–reality relationship, 95
　materialist perspective of, 42–43
　structuralism influenced by, 120
Marxism
　Althusser's structuralist view of, 118–119
　critical analytic strategies in, 17
　growth in 20th century, 116
　ideas–reality relationship in, 95–96
　material cause, 27
materialism, 8, 34, 96, 153*b*
matter, as agentic, 148, 150
Maxwell, J., 60–61
Mazzei, L., 147–148
Mead, G. H., 105, 126
meaning
　critical realist views of, 61
　deconstruction of, 160
　made *vs.* discovered, 3
　vs. nihilism, 3
　post-structuralist views of, 119, 121, 123*b*, 160
　in qualitative research practice, 5
　social practice connected to, 74
　structuralist views of language and, 117–118, 121, 137
　symbolic interactionist views of, 105–106
meaning condensation, 83
medieval philosophy, 29–31, 32
Meditations on First Philosophy (Descartes), 35–36
Menand, Louis, 92, 96
mental representations, 35
Merleau-Ponty, Maurice, 11, 69, 71, 72, 88
meta-ethnography, 16
metaphors, analytical, 107
metaphysical stage of knowledge, 49
metaphysics. *See also* ontology
　defined, 8
　in defining qualitative research, 7
　in Greek philosophy, 22–23, 28
　positivist critique of, 18
　as theoretical subdiscipline of philosophy, 7
metaphysics of presence, 120, 152
Michell, Joel, 54

microsociology, 108, 113
Mill, John Stuart, 51–52
Mills, C. Wright, 79
mind, philosophical paradigms for, 37–38, 45, 149–150
mind/soul–body relationship, 24–25, 36, 39, 126
"mirror of nature," 45, 149–150
modernity, Foucault's critique of, 125
modern philosophy
  epistemological approach of, 36, 39, 43, 44, 45
  ideas as viewed in, 35, 36, 37, 38, 39, 43, 45
  key figures
    Berkeley, 37–38
    Descartes, 34–37, 39, 93–94
    Hegel, 41–43, 120
    Hume, 38–40, 48–49, 157
    Kant, 40–41, 120
    Kierkegaard, 43
    Locke, 37, 38
    Marx, 17, 42–43, 58, 95, 120
    Moore, 40
    Nietzsche, 18, 43, 120, 148
  natural science as influence on, 34–35, 38–39, 40, 41, 43–44
  objective–subjective split in, 21, 28, 41, 44
Moghaddam, F. M., 134–135
Montaigne, Michel de, 34
Moore, G. E., 40
Moralität (Kant), 42
morality
  as element in all sciences, 102
  ethical life related to, 42
  in modern power relations, 125
  in positioning theory, 134, 135
  science of *la morale*, 50
  universal reason as source of, 41
  as unknowable, 39, 41
moral philosophy. *See* ethics
"moral science," 51, 52, 102
Morgan, D. L., 113, 114
myth of the given, 152

N

naturalism, 50–51, 69
naturalistic fallacy, 40
natural laws, 33, 98
natural sciences
  epistemology prompted by, 10
  hermeneutical views of, 76, 78, 79
  materialism in, 8
  modern philosophy influenced by, 34–35, 38–39, 40, 41, 43–44
  objective–subjective split driven by, 44, 155–156
natural units, *vs.* central themes, 83
nature, as agentic, 148
nature–culture dualism, 142, 148, 154
neopragmatism. *See* Putnam, Hilary; Rorty, Richard
Neurath, Otto, 52–54
new materialism, 153*b*
"new philosophies" in qualitative research. *See* new materialism; post-qualitative research
Newton, Isaac, 29, 38–39
Nichomachean Ethics (Aristotle), 13, 25, 26
Nietzsche, Friedrich, 18, 43, 120, 148
nihilism, 3
Noblit, G. W., 16, 48
nominalism, 30–31
Nothing Is Hidden (Malcolm), 74–75
"nothing is hidden" theme
  in phenomenology, 53, 74–75
  in positivism, 53
  in pragmatism, 18–19, 108, 159
nous, 15

O

objective–subjective split
  in modern philosophy, 21, 28, 41, 44, 155–156
  post-qualitative rejection of, 147
objectivism, critique of, 160
object–subject dualism, 36, 72
observation, theory related to, 57
obvious, making dubious
  as analytic strategy, 16, 18
  in postmodernism, 130, 137

in post-structuralism, 18, 116, 123–124*b*, 137, 160
obvious, making obvious
 as analytic strategy, 16, 17–18
 in hermeneutics, 17–18, 75, 88
 in phenomenology, 17–18, 75, 82, 84–86*b*, 88, 108, 159
 in pragmatism, 108
Ockham's razor, 31
Olesen, V., 141
ontic logos, 155
ontological hermeneutics, 76
ontology. *See also* metaphysics
 in critical realism, 58–59, 60, 157
 defined, 8
 in defining qualitative research, 7
 flattened, 149
 fundamental, 76
 new materialist, 153*b*
 post-qualitative deconstruction of, 147, 150, 153*b*, 154
 pragmatic approaches to, 159
 in qualitative research, 9
 realism *vs.* anti-realism in, 8–9
 realist approaches to, 60
 as theoretical subdiscipline of philosophy, 7
Oration on the Dignity of Man (Pico), 32
Order of Things, The (Foucault), 127

P

paideia, 2
panopticon, 126
paradigms, 57, 58
paradigm shifts, 57
Park, Robert, 104–105
Parker, Ian, 131
Parmenides, 22, 23
parole, 117
Passions of the Soul, The (Descartes), 36
Peirce, Charles Sanders, 92, 97, 98, 100, 102
Pelias, R., 17
perception, philosophical views on, 37–38, 92–93
Perception and Discovery (Hanson, et. al), 57
performativity, 129, 135–136
permanence, *vs.* change, 22–23
perspective, invention of, 32
perspectivism, 32–33, 120
Phaedo (dialogue), 4
phenomenalism, 51, 55, 56
phenomenology
 analysis in, 164
 anxiety study, 84–86*b*
 clarification as goal of, 67–68
 defined, 66, 80, 159
 as descriptive enterprise, 67, 79
 as eidetic, 68–69
 essentialist approaches to, 68
 existential. *See* phenomenology, existential
 experience described by, 65–66, 67, 68, 69, 152
 as German tradition, 17–18, 66, 91
 goals of, 65–66
 vs. hermeneutics, 88, 159
 as investigation of essences, 80, 84*b*
 key figures
  Aristotle, 25
  Husserl, 51, 66–69, 71, 72, 79, 80, 88
  See also under phenomenology, existential
 obvious made obvious in, 17–18, 75, 82, 84–86*b*, 88, 108, 159
 vs. positivism, 67
 positivist influence on, 51, 157
 post-structuralism influenced by, 152
 pragmatism as American version of, 88
 in qualitative research, 80–83, 89, 161
 as realist approach, 71–73
 reduction as methodology in, 80–83, 85*b*, 89
 as reflective inquiry, 69
 "Throwing Like a Girl" study, 81–83

phenomenology, existential
　existential conditions for
　　experience, 69, 72
　key figures
　　Heidegger, 69, 70, 71, 72, 76, 79, 88, 120, 159
　　Merleau-Ponty, 11, 69, 71, 72, 79, 88
　　Sartre, 43, 73–74, 115, 116
　　Wittgenstein, 11, 19, 74–75, 109
　pragmatism related to, 92
　scientific practice as understood by, 72–73
　structuralism's challenge to, 115
Philosophical Investigations (Wittgenstein), 74
philosophy. *See also* qualitative research, philosophies of
　basic questions of, 7–16
　choosing and engaging, 161–162
　defined, 1, 4
　Greek. *See* Greek philosophy
　human finitude linked to, 4
　life as focus of, 4–5
　modern. *See* modern philosophy
　origins of, 2–4
　political, 7, 12–16
　practical, 7, 21
　　See also ethics; hermeneutics; political philosophy
　as practical foundation of qualitative research, 162–165
　public, 15
　Renaissance, 31–34
　theoretical, 7, 21
　　See also epistemology; metaphysics; ontology
　as way of life, 4–5, 64, 89, 161–162
philosophy, Western, history of
　Critchley on, 3–4
　current vs early conceptions, 2
　deconstruction, 120
　Greek philosophy. *See* Greek philosophy
　materialism *vs.* idealism in, 8
　medieval philosophy, 29–31
　modern philosophy. *See* modern philosophy
　Plato and Socrates, 2–3
　realism *vs.* anti-realism in, 9
　Renaissance philosophy, 31–34
　20th-century philosophy. *See* 20th-century philosophy
Philosophy as a Way of Life (Hadot), 2
phronesis/phronimos. *See also praxis*
　in ethics of reporting, 165
　as *praxis*-related virtue, 13, 14, 25–26
　in qualitative research, 15
　social sciences as, 13, 15, 79
Pico della MIrandola, Giovanni, 32
Plato
　body–soul duality, 24, 96
　cave allegory, 23, 44–45, 93
　Christianity synthesized with philosophy of, 24, 29
　contemporary reprisals of, 31
　cosmological context for ideas, 35
　dialogues with Socrates, 2–3
　in history of philosophy, 23–24
　metaphysical theory, 23, 28, 43
　on problem of universals, 30
　visual metaphors, 45
Platonism, 93
poetry, as analytic stance, 17
poiesis, 14. *See also techne*
political philosophy, 7, 12–16
polycentric cosmology, 32
Popper, Karl, 55–56, 58
position, 135
positioning theory, 134–136
positioning triangle, 135
positivism
　as anti-realism, 48, 56, 61–63
　beaurocratic, 55
　as British tradition, 48, 51, 53, 64, 91
　causality in, 39, 49, 56, 63, 157
　classical, 48, 49, 52, 54, 55
　critiques of, 54–55, 57–58
　data collection/analysis in, 164
　defined, 157
　influence of, 50–51

key figures
  Comte, 48, 49–50, 52, 54
  Hume, 28, 38–40, 41, 48–49
  Mill, 51–52
  Neurath, 52–54
  Vienna Circle, 52
knowledge as experience-based, 49, 157
metaphysics critiqued by, 18
methodological focus of, 47–48, 60
mischaracterized in qualitative research, 48, 54–55, 61–63
vs. phenomenology, 67, 157
postmodernist similarities to, 51, 53
principle of verification, 53–54, 55, 157
in qualitative research, 54–55, 59–60, 63–64, 157
vs. realism, 48, 55, 56, 57–58, 64, 156
values inherent in, 11–12
Vienna Circle approaches to, 52–53, 55
posthumanism, 127, 147, 150
Postmodern Condition, The (Lyotard), 129
postmodernism
  artistic representation of, 32–33
  key figures
    Baudrillard, 129–130
    Lyotard, 129
  objectivism criticized in, 160
  obvious made dubious in, 130, 137
  positivist similarities to, 51, 53
  structuralist influence on, 137
post-qualitative research
  data collection in, 164
  diversity within, 146–148
  humanist methodology rejected by, 151
  individual knowledge questioned by, 146
  ontology deconstructed in, 147, 150, 153*b*, 154
  philosophy guiding, 122, 147–150
post-structuralism
  data collection/analysis in, 164
  deconstruction as analytic strategy in, 119, 160
  defined, 119
  key figures
    Derrida, 18, 70, 120–122, 136
    Foucault, 43, 122–129, 130–134, 136
  objectivism criticized in, 160
  obvious made dubious in, 18, 116, 123–124*b*, 137, 160
  positioning theory, 134–136
  postmodernism. *See* postmodernism
  in qualitative research, 123–124*b*, 130–134, 162
Potter, J., 152–153
power
  acknowledged in discourse analysis, 130–131, 133
  philosophical paradigms for, 43, 125–126
practical inquiry, *vs.* theoretical inquiry, 13.21
practical sciences, 13
practice, primacy of, 102
pragmatism
  abduction as core of, 103–104, 110, 113
  as American tradition, 21, 91, 105, 159
  anti-realism *vs.* realism in, 113
  as bridge between analytic and Continental philosophy, 102–103
  vs. Cartesianism, 94
  defined, 159
  epistemology as viewed in, 10–11
  experience as viewed by, 93–94, 99, 113–114
  global recognition of, 91–92
  hidden made dubious in, 18–19, 111–112*b*, 113, 159
  historical development of, 21
  ideas as viewed in, 92–96, 99–100
  inquiry as life process, 100
  key figures

pragmatism (cont.)
Dewey. See Dewey, John
Holmes, 92
James, 92, 97–98, 100, 101, 102
Peirce, 92, 97, 98, 100, 102
Putnam, 12, 101–102
Rorty, 101, 102, 103
knowledge as viewed in, 10, 91, 100, 106, 111–112b, 114
vs. Marxism, 95–96
"nothing is hidden" as theme of, 18–19, 108, 159
obvious made obvious in, 108
ontology of, 159
vs. Platonic philosophy, 93
practical reasoning as focus of, 104
in qualitative research, 100, 103–110, 111–112b, 113, 114, 162–163
reflexivity acknowledged by, 95–96
as set of theses, 101–102
as shift in perception, 92–93
truth as viewed in, 96, 97–98, 111b
praxis, 14, 15, 25–26. See also phronesis/phronimos
prejudices, 76, 77, 78, 87
preparation, philosophical influence on, 163
presence, metaphysics of, 120, 152
"presence-at-hand," 71, 72
pre-understanding, 77
primary qualities, vs. secondary qualities, 37
principle of verification, 53–54, 55
Protagoras, 53
psychologism, 67–68
psychology
descriptive, 76
discursive, 132–133b, 134
philosophical influences on, 40, 52
public philosophy, 15
Putnam, Hilary, 12, 101–102

Q

qualia, vs. quanta, 44
qualitative research, methods and techniques
analytic strategies, 16–19
CAQDAS, 60, 63–64

conversation analysis, 131
discourse analysis, 130–134
feminist influences on, 141–142
hermeneutic influences on, 78–79, 80, 83–88, 161
indigenous, 145–146, 154
interviews, 15, 62b, 123–124b
mixed-methods approaches, 114
phenomenologist influences on, 80–83, 89, 161
positioning theory, 134–136
positivist influences on, 54–55, 59–60, 63–64, 157
post-structuralist influences on, 123–124b, 130–134, 162
pragmatist influences on, 100, 103–110, 111–112b, 113, 114, 162–163
racism as factor in, 9, 132–133b, 144
realist influences on, 60–61, 62b
research phases, 163–165
structuralist influences on, 130, 162
qualitative research, philosophies of
American. See pragmatism
British. See positivism; realism
"choosing" and engaging philosophies, 161–162
compared, 158t
feminist, 141–142, 154
French. See post-structuralism; structuralism
German See hermeneutics; phenomenology
indigenous, 142–146, 154
modern humanist underpinnings, 146–147
new ontologies/new materialisms, 153b, 160
paradigmatic controversies and contradictions in, 156–157
practical application of, 162–165
qualititative research
"death" of, 127
defined, 5–6
epistemology as problem for, 7, 11
forms of knowledge related to, 15

shaped by subjective-objective
    split, 21, 28, 44, 156
  in social sciences, 103–110
  values underlying, 12, 40
quanta, vs. qualia, 44
"quantitative imperative," 54
Quine, W. V., 57, 58

R

racism, 9, 132–133*b*, 144
randomized controlled trials, 111*b*
Raphael, 25
rationalism, critical, 56
rationality, philosophical paradigms
    for, 27, 33, 40
Ray, L., 145
"ready-to-hand" mode, 71, 72
real domain, 58–59
realism
  Adorno study, 62–63*b*
  analytic strategies in, 17
  as answer to problem of
    universals, 30
  vs. anti-realism, 8–9, 10, 31, 56, 64,
    113, 155, 164
  as British tradition, 17, 64, 91
  causality in, 56, 61
  critical *See* critical realism
  data collection/analysis in, 164
  existential phenomenological
    outlook as, 71–73
  falsification as mark of
    scientificity, 56
  hidden made obvious in, 56,
    62–63*b*, 64, 159
  key figures
    Bhaskar, 58
    Hanson, 57, 58
    Kuhn, 57–58
    Mawell, 60–61
    Popper, 55–56, 58
    Quine, 57, 58
  in qualitative research, 60–61, 62–63*b*
  as response to positivism, 48,
    55, 57–58
  validity criteria, 61
reality, as *ontic logos,* 155
reason, 21, 26–27, 104

reasonableness, *vs.* rationality, 33
reduction, phenomenological, 80–83,
    85*b*, 89
reflexivity, 95–96, 109, 151. *See also*
    triple hermeneutics
Renaissance philosophy, 31–34
reporting, philosophical influences
    on, 164–165
representationalism
  deconstructed in post-qualitative
    research, 150, 151
  phenomenology's break with, 69
  philosophical presuppositions,
    6, 7, 35
  in research practices, 151, 164
Republic, The (Plato), 24
res cogitans, 35, 36
researchers, philosophical
    presuppositions about, 5, 7
research ethics, 11. *See also* ethics
res extensa, 35, 37
Richardson, F. C., 78
Ricoeur, Paul, 77, 118
Robinson, D. N., 36
Rorty, Richard, 45, 101, 103, 149–150
rules, in positioning theory, 134

S

St. Pierre, Elizabeth, 147–149, 150,
    151, 152
Sartre, Jean-Paul, 43, 73–74,
    115, 116
Saussure, Ferdinand de, 115, 116–117,
    136, 137
Schleiermacher, Friedrich, 75, 88
School of Athens, The (Raphael), 25
Schrift, A., 119–120
Schütz, Alfred, 108
Schwandt, T., 119
science. *See also* natural sciences;
    social sciences
  cognitive, 35
  critical nature of, 17, 58
  descriptive *vs.* explanatory
    tasks of, 65
  morality in, 50, 102
  social *vs.* physical, 65, 155–156
Science Wars, 8

scientific knowledge
  critical realist approaches to, 47–48
  positivist/inductive/verification approaches to, 47, 157
  realist/deductive/falsification approaches to, 56
  subjectivity of, 57
  value-neutrality of, 11–12, 144
Scientific World Conception, The (Neurath), 52
secondary qualities, vs. primary qualities, 37
Second Sex, The (Beauvoir), 140
self, philosophical paradigms for, 38, 128–129
semiotics, 98
sensitizing concepts, 87, 106, 134
Shakespeare, William, 34
signs, 98, 129–130
Simmel, Georg, 107
social constructivism, 8, 28, 150
social order, 108–110, 134, 135
social reform, 105
social sciences
  interpretation in, 9
  as moral and political enterprise, 13
  as *phronesis*, 13, 15, 79
  vs. physical sciences, 65, 155–156
  pragmatism in, 103–110
  as public philosophy, 15
  rooted in practical sciences, 33
  symbolic interactionism in, 105–106
  values in, 11–12
sociology
  microsociology, 108, 113
  origins of, 104
  philosophical influences on, 50–51, 104–105
Socrates, 2–3, 4
sophia, 15
soul, philosophical paradigms for, 24–25, 36, 39, 45, 126
soul/mind–body relationship, 24–25, 36, 39, 126
storylines, in positioning theory, 135
Straus, Erwin, 81

structuralism. *See also* post-structuralism
  analysis in, 164
  defined, 160
  key figures
    Althusser, 116, 118–119
    Lévi-Strauss, 115–116, 118, 120, 136
    Saussure, 115, 116–117, 136, 137
  in qualitative research, 130, 162
Structure of Scientific Revolutions, The (Kuhn), 57
Studies in Ethnomethodology (Garfinkel), 108
subjectification, 125–126, 130, 134
subject–object dualism, 36, 72
subject positions, 133, 134, 151
Sullivan, W. M., 15
Swidler, A., 15
symbolic interactionism, 105–110
systematic doubt, method of, 35–36
System of Logic, A (Mill), 51

T

tabula rasa, mind as, 37
Taylor, C.
  humans as self-interpreting animals, 77, 78
  on overcoming epistemology, 10–11
  phenomenological approach, 71, 72–73
  Sartre criticized by, 73
  understanding as condition of being human, 76
  on validity of human and social theory, 78
techne, 14, 15, 165. *See also poiesis*
technologies, in discourse analysis, 133
technologies of the self, 127–128
teleological worldview, 28–29, 44
text, post-structuralist views of, 120–121, 122
textual empiricism, 131
Thales, 22
Theaetetus (dialogue), 2–3
theocentric cosmology, 32, 35

theological stage of knowledge, 49
theoretical inquiry, *vs.* practical inquiry, 13, 21
theoretical knowledge. *See theoria*
theoria, 14, 59, 165. *See also episteme*
theory, as practice, 57, 93
Thomas, William, 104
Thomas Aquinas, 29, 30
"Throwing Like a Girl" (study), 81–83
Tipton, S., 15
Toulmin, Stephen, 33–34, 92
Treatise of Human Nature (Hume), 38
triple hermeneutics, 80, 95. *See also* reflexivity
truth
  correspondence theory of, 111*b*
  pragmatic views of, 96, 97–98, 101, 111*b*
Truth and Method (Gadamer), 79
20th-century philosophy
  American. *See* pragmatism
  British. *See* positivism; realism
  French. *See* post-structuralism; structuralism
  German. *See* hermeneutics; phenomenology

U

understanding, as existential, 70
universalism, 31, 144
universals, medieval problem of, 29–30

V

values
  in critique of positivism, 54
  fact–value dichotomy, 12, 102
  as non-natural, 40
  as "oughts" in is/ought dichotomy, 28, 39–40
  in qualitative research practices, 12, 40
  in science of *la morale*, 50
  social reasoning about, 15
  in social sciences, 11–12
  as subjective, 39, 44
van Kaam, A., 80
verificationism, 53–54, 55, 157
Vienna Circle, 52
virtues of knowledge, 14, 14*t*, 25–26

W

Walkerdine, V., 130, 131
We Have Never Been Modern (Latour), 148
Whitehead, Alfred North, 24
white swan analogy, 55–56
William of Ockham, 31
Wittgenstein, Ludwig, 11, 19, 74–75, 109
wonder, as beginning of philosophy, 2–3
world, philosophical paradigms for, 5–6, 7, 64, 71
Wrathall, M., 71

Y

Young, Iris Marion, 81–83

Z

Žižek, Slavoj, 119
Znaniecki, Florian, 104
zoon politikon (political animal), 12, 27